NEW APPROACHES TO POLYMER MATERIALS

NEW APPROACHES TO POLYMER MATERIALS

G.E. ZAIKOV
Editor

Nova Science Publishers, Inc.
New York

Art Director: Maria Ester Hawrys
Assistant Director: Elenor Kallberg
Graphics: Denise Dieterich, Kerri Pfister,
Erika Cassutti and Barbara Minerd
Manuscript Coordinator: Sharyn Schweidel
Book Production: Tammy Sauter, and Benjamin Fung
Circulation: Irene Kwartiroff and Annette Hellinger

Library of Congress Cataloging-in-Publication Data

New approaches to polymer materials / G.E. Zaikov (ed.)
p. cm.
Includes bibliographical references and index.
ISBN 1-56072-254-1 :
1. Polymers. I. Zaikov, Gennadii Efremovich.
TA455.P58N44 1995 95-24680
620.1'92--dc20 CIP

6080 Jericho Turnpike, Suite 207
Commack, New York 11725
Tele. 516-499-3103 Fax 516-499-3146
E Mail Novasci1@aol.com

Printed in the United States of America

CONTENTS

INTRODUCTION

Over 130 millions of tons of artificial polymers and approximately the same quantity of natural polymers are produced now in the world (cellulose and its derivatives, natural rubber and so on). Together they make up over 250-260 million tons. If this quantity is converted to volume, it gives 240 million cubic metres of polymers per year. We can mention the following figures to compare: there are approximately 100 million cubic metres of steel and 14 millions of cubic metres of non-ferrous metals produced per year. In other words, the same volume of polymers on the one hand and steel, cast iron, rolled metal and non-ferrous metals in aggregate on the other hand, are produced now in the world.

But this is not all. The dynamics of development are also important. Here we have the following figures: During the period from 1995 to 2010 the volume of polymer production will increase by 2 times while the increase of production for the basic metals will increase only 1.5 times.

ANALYSIS OF RECENT STATE IN THE PRODUCTION AND THE CONSUMPTION OF POLYMERIC MATERIALS IN RUSSIA

R.M. Aseeva, G.E. Zaikov
Institute of Chemical Physics, Russian Academy of Sciences, 4 Kosygin Str., 117334 Moscow, Russia

Abstract — In the paper the recent state and the development of polymer industry in Russia have been considered. The production structure of principal sources of raw material and energy have been discussed. The dynamics in the production and the consumption of the most important kinds of polymeric products have been reviewed.

Polymeric materials play a very important role in the modern society. Raw material is a defining factor for economics of the production of synthetic polymer materials since specific weight of raw material costs in general expenditures for polymer production usually achieves to 70-80% and in some cases to 90%. At the present time, processing products of oil, natural gas or gas oil condensates and coal are principal raw material bases for the production of resins and plastics, chemical filaments and rubbers, other kinds of polymer materials. Table 1 illustrates the dynamics in recent production of energy and raw material resources in Russia. There are used the conventional fuel values for the comparison of different kinds of energetic sources.

Table 1. The production of energy resources in Russia [1-3] (million tons of conventional fuel)

	Years						
Product 1985	**1990**	**1991**	**1992**	**1993**		**1994/95 x**	**2010x**
coal	273	273	241	231	205	185/180.1	228-241
oil	775	738	659	570	493	458	–
gas	536	743	746	773	716	720	1147
electric energy	48	49	49	49	51	–	–
in whole	1690	1859	1752	1649	1521	–	–

x - estimation

Russia possess by about 20% of world supplies of coal and more 40% of natural gas ones. The main supplies of energetic resources are concentrated in the regions of Siberia, the Ural, Far East and to a less extent at European part of Russia. As an example, in 1993 , 70% of the total coal was obtained by Siberia and the Ural regions, 13.1% of one by Far East and 16.9% of all coal were obtained at European part of Russia [1]. The Russia supplies of chemical raw material and energy averaged about 75% of the former USSR potential. 90% of oil output of former USSR are carried out by Russia [4].

A decrease of coal output and particularly catastrophical downfall of oil output are observed (Table 1). By 1995, the volume of oil output is expected to achieve to own consumption one [3]. There are relatively stable situation for natural gas output, which is characterized by 5% annual growth [5].

The great expenditures on the output and transport of coal, gas and oil are connected with the complicated geological conditions and unfavorable accommodation of many working deposits as regards main consumers of energy resources both on internal market and foreign one. The continuous growth of the cost of energetical and raw material resources is the reflection of general social processes, taking place due to the degradation of former USSR. The analysis shows that the most rapid growth of the prices was observed in the period of 1992-1993 years: oil cost rises in 14 times, natural gas - in 70 times, coal - in 72 times and electrical energy - in 69 times [2]. The prices continued to rise in 1994 (Figure 1). The tariffs on railway transport only in 1993 increased in 21 times [1]. As result, at the average distances about of 1300km more half of coal coast, as example, is formed by transport expenses.

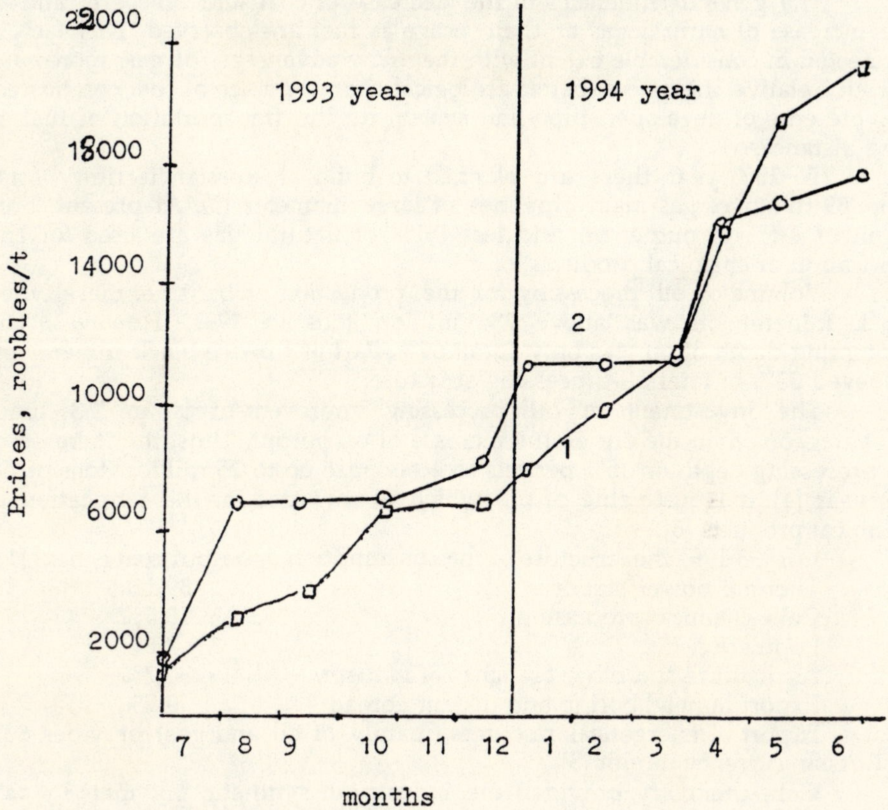

Figure 1. The increase dynamics of average prices on coal (1) and its transportation (2).

Gas, oil (black mineral oil) and coal are a principal fuel source but at last time their specific weight in fuel balance of Russia is considerably changed (Table 2).

Table 2. Specific weight of different energy sources in fuel balance of Russia (%) [1]

Fuel	Years			
	1985	1990	1992	1993
gas	44.7	59.0	61.4	62.0
oil (black mineral oil)	21.9	12.1	12.2	11
coal	33.4	28.9	26.4	26

The general tendencies in the decrease of coal and oil parts and in the increase of natural gas at their usage as fuel are observed. This fact is connected in considerable extent with the many advantages of gas: more easy output, relative cheapness, better ecological characteristics of fuel combusted, the presence of developed pipe-line system for the transportation of fuel at long distances.

To 2000 year there are planned to build at Russian territory some more 69 th.km of gas main pipelines of large diameter [5]. At present, only about of 4-5% of output oil and less 0.1% of natural gas are used for the fabrication of chemical products [6].

Volume of oil processing for the production of benzine, diesel fuel, black mineral oil was above 284 million tons in 1991. However, the processing depth of oil was only about 65%. Part of furnace black mineral oil achieved 33% in total oil-processing structure.

The investment in oil processing improvement is in 2-3 times advantageous than the one in the increase of oil output. Thus, the increase of oil processing depth on 15% permits to economize up to 25 million tons of oil per year [4]. It is just value of oil, which is now used for the fabrication of chemical products [6].

In 1993-1994 the structure of the consumption of output coal is next [1]:

Thermal power station	39.1%
Coke-chemical processing	13.5%
Population	7.6%
Agriculture and other branches of industry	30.7%
Export in neighboring and distant abroad	9.7%

Export of energetical resources (mainly of oil and gas) provides 60% of Russian currency receipt [5].

Coke-chemistry provided the industry of synthetic polymers by raw material up to 80% in 1975. By 1980 its part in raw material balance of polymer products was decreased to 60%, by 1985 - to 50% and by 1990 - to 40-45%.

The problem of chemical processing of coal as important source of hydrocarbon raw material is very actual because of the exhaustion of oil resources in the near future. The intensive fundamental investigations are carried out towards effective usage of chemical potential of coal. A new thermo-chemical technologies for coal processing in liquid and gaseous fuel, mining wax, dispersive carbon for the production of plastics and resins, heat resistant ceramics, etc. have been developed [7].

The changes in the structure of expenditures for the production of chemical products were observed because of economic situation in the country. In 1993, specific weight of raw material in the total expenditures on the fabrication of chemical products was decreased from 59% to 49% but specific weight fuel or energy was increased from 10.5% to 19.3% in comparison with 1991 [2]. The important problem is the improvement in the resource cycle of the fabrication of chemical products: the using of low or unwasteful technologies, of energy saving ones and processing of wastes or secondary raw material [8]. Ecological aspects of the production and the application of polymer materials are of great significance.

Russian chemical industry are a complicated many-branch complex. Russia has always occupied a leading position in industrial potential of former USSR. An average of about 67% of total output of USSR goods production were carried out by Russia, including the output of [6]:

	(in%)		(in%)
polymers	-74	synthetic rubbers	-80
plastic articles	-80	chemical filaments	-<60
synthetic dyes	-75		
products of heavy organic synthesis	-76		
Chemicals and particularly pure substances	-90		
mineral fertilizers	-65		

It should be noted that specific weight of chemical production (without chemico-pharmaceutical one) in total volume of national product of Russia is only 6.9% [6].

In period of 1975-1985 USSR chemical industry were developed with high rate of growth: the volume of its production increased in 2.5 times at double growth of industrial one on the whole [9]. The output of chemical products begin to fall in the end of 80th. If in 1985-1988 the annual growth of the production was still equal 3.5-5.5%, in 1989 the increase became equal only 0.6% and then the reduction of one was observed. In 1990 the downfall exceeded 2%, in 1991 - 6%. In 1992 the tendency was intensified. The downfall of chemical production achieved to 20%. In comparison with 1991 year there are observed the reduction of the output of next polymers: homo- and copolymers of styrene - on 37%; polypropylene - on 16.3%, polyethylene - on 5.4%, polyvynilchloride - on 14%, chemical fibers - on 10% [10].

In 1993 the volume of the fabrication of chemical products was decreased on the whole on 22% if compare with one in 1992 year. It should be emphasized that the most downfall was observed for the output of principal, basis polymeric products: synthetic rubbers and latexes (31.6%), chemical fibers and threads (26%%), varnish-paint materials (30%), polypropylene (22%). Only volume of the production of polyethylene was retained at the level of one for 1992 year [2].

Table 3 shows the dynamics of recent production of principal polymeric materials in Russia.

Table 3. Dynamics of the production of polymeric materials in Russia [11,12] (Thousand tons)

Product	Years						
	1985	1988	1990	1991	1992	1993	1995^x
Resins and plastics	(6000^x) 4115	4490	(6700^x) 4430	4160	3210	>2500	–
Chemical filaments	1600	–	(1850^x) 673.2	528.7	405.0	299.7	–
Synthetic rubbers and latexes	2020	2232	(2700^x) 2157.7	2012.3	(1928^x) 1610.8	1379	1465

x - planned output

The earlier declared need in resins and plastics on 1990 year (10 million tons [13]) was proved to be in 2.5 times more than actual value of their output. However, the sharp downwall of the production are typical practically for all branches of Russian industry (machine-building, agriculture, food, textile and other ones). As result, in 1992-1993 the volume of realizated polymeric goods was only 71-72% of total production of polymeric materials [2].

At present time, the general strategy in the development of polymer manufacture in Russia are oriented towards accelerated growth of progressive kinds of polymers, the reduction of assortment of polymers produced by obsolete technologies. The important direction is the creation both new technologies and new polymeric materials for the usage in nontraditional areas.

In period of 1980-1993, the thermoplastics part increased and thermoreactive polymer one considerably decreased in the structure of the production of plastics. Synthetic fibers became to occupy the leading position in the output of chemical fibers. The manufacture of synthetic rubbers and latexes of special fixing developed with outstrip rate [4].

Let's consider in detail the situation in the production structure of the most important polymeric materials.

Polyolefines, polymers of vynilchloride and styrene are occupied the leading position among polymerizational thermoplastics.

Polyolefine industry in Russia is the most large-scale one. PO are occupied among other plastics the first place by the value of their production. Polyolefine part in total output of resins and plastics achieve to 24.3% [13]. The capacities for the manufacturing of PO are equal to 1.53 million tons/year including 940 th.t/year one for PEHP, 450 th.t./year one for PELP and 140 th.t./year one for PP [14]. There were supposed to increase the production capacities for PO in period of 1990-1995 years as by the introduction of new factories up to 1990 (as example, for PE- at IA "Tomskoilchim", for PP - at IA "Stavropolpolymer") as by intensification and technical reequipment of acting ones (as example, at IA "Orgsynthesis" in Kazan to increase the capacity on 56.5 th.t/year) [14]. There are the prospect in the creation of 3 oil-chemical complexes (in Tobolsk, Surgut, New Urengoy) for polymer production, including PO one, on basis of hydrocarbon raw material of West-Siberian deposit. However, now the building of OCC is practically freezed because of insufficient budget foundation and the absence of native or foreign investments [10].

It should be noted that the acting production capacities up to 1990 year were used on 75-85%. In period of deep economic crisis the using coefficient of the production capacities was reduced to 60 -30% [2]. The level in the satisfaction of the PO demand does not now exceed 60%.

In spite of crisis situation, in Russia there are carried out the works towards the improvements of all 4 methods for the production of polyolefines: gas-phase, suspensional ones, in solution and by high pressure technology. Due to the using of active and selective catalysts the reduction of some steps in technological process and also the improvements of the quality of PO have been achieved [15-17].

The technologies for the production of linear PE of low density and superlow density by copolymerization of ethylene with higher α-olefines and also the one for new polymer of 4-methylpentene-1 ("Templene" trademark) have been developed [17]. Templene may be used as dielectric material in radio-electronics or as constructional one in machine-building.

In 1988 year "Tekhnokhim" Interbranch State Association was created for the production development of PO, PVC, styrene polymers, PVA, fluorinated polymers, thermoset plastics and different composite materials.

The second place among thermoplastics is occupied by PVC and VC copolymers. Their part in a total volume of polymer production is 18-20%. One of main consumers of thermoplastics is cable industry. In total volume of polymer consumption by cable industry PO part is 25-26% and PVC plasticates one is 62-65% [18]. Russian cable industry offer a great assortment of polymeric materials: more 20 trademark of PVC-plasticates, up to 30 compositions based on PE of low and high density, 6 trademark of fluoroplastics, films based on polyimides, polyethyleneterephtalates, polyamides and others.

The 3rd large scale production of polymers is the one of polystyrene and styrene copolymers, as example, ABS. Their part achieves to 9% [13]. The synthesis of plastics based on styrene in Russia is carried out both by periodic ways - emulsive, suspensive or block-suspensive polymerization and by continuous method of polymerization in block mass. At present time, above 65% of high impact polystyrenes are produced by the continuous method, which permits to use automatics in technological process and to mechanize almost all stages of the one. It should be noted that the production capacities for styrene polymers have no increase in period of 1985-1994. The extent of the saturation of PS market is equal 60% [13].

The significant part of polymers produced in Russia are used for the fabrication of varnish-paint materials (VPM). In 1990 year the demand on VPM was estimated at level of 3500th. t. However, market saturation by VPM was only 64.5%. Deficit was partially filled up by VPM import at 160th.t. (it is 7.3% of native production). The import of VPM in 1993 was increased to 170th.t. [19]. The structure of the production of VPM and dynamics of the development are shown in Table 4. Total downfall of VPM production was observed. However, in period of 1990-1993 the part of traditional products (oil-paints, drying-oils, products based on cellulose esters) was decreased from 53 to 42.8%. At the same time the part of progressive kinds of VPM was increased from 47 to 57.2%. A great popularity have the water-soluble, without solvents and powder VPM based on epoxydes or epoxy-polyester hybride systems, produced by JC "Pigment" (Sanct-Peterburgh). In present time the development of VPM assortment in significant extent depend on the execution of ecological safety requirements for the products.

Table 4. The volume and the production structure of varnishes and paints (thousand tons) [18]

	Years		
Assortment	**1990**	**1992**	**1993**
Varnishes on condensation resins,	483.8	286.9	204.9
including semi-products	227.0	212.0	168.8
Enamels on condensation resins	417.0	270.3	215.2
Varnishes on:	67.8	40.8	25.1
polymerization resins			
cellulose esters	300.6	173.9	122.4
Diluters	142.1	83.7	55.8
Water-emulsion paints	59.7	37.1	27.4
Oil-paints	458.99	194.4	113.0
Drying oils,	378.2	147.9	80.2
including semi-products	162.0	69.8	40.9
Powder paints	2.8	1.2	1.4
Others (alcohol varnishes, thermoplastics)	19.2	5.1	5.7
On the whole,	2330.1	1241.3	851.1
without semi-products	1841.1	959.5	641.4

In the production and the consumption of VPM acrylic polymers occupy the 3rd place after alkyde resins and polyesters. Their part in the total consumption structure for VPM exceeds 20%. The coatings based on acrylic VPM are characterized by high atmosphere - and light resistance, by the resistance to the action of water and alkaline solutions. Russian market forming for acrylic raw material, including VPM one, is carried out in the main by JC "Acrylate" (Dzerzinsk, Novgorod region) [20].

Modern techniques stimulate the development of new constructional and functional polymeric composite materials (PCM). The fast increase of PCM production in Russia has been associated with re-orientation on the consumption growth of non-metallic constructional materials. Volume of PCM production must be increased from 2th.t (1988 year) to 60th.t. (1995) and to

200th.t. Flexible, semi-rigid and rigid foams, with many variations for specific applications, are above 65% of real, total PU consumption in Russia [22,23].

The large-scale production of chemical fibers include 5 groups of polymers: viscose, cellulose esters, polyesters, polyamides, PAN and PO fibers. At present time, the consumption of natural and viscose fibers is almost equal to the one of synthetic polymeric filaments. The main direction in the development of synthetic polymer fibers for mode of life is the approach their properties to one of natural filaments. PAN fibers have the most chance to be substituted of full value for natural filaments.

There are progress in the production of fibers for technical fixing, in the main due to the increase of its assortment: fibers with high strength and high module of elasticity, thermal and fire resistant ones, fibers with special properties.

Dynamics in production development of synthetic rubbers and latexes is shown in Table 5. In structure of production of synthetic rubbers in 1991-1992 rubbers of general fixing are predominated over ones of special fixing (84% and 11% correspondingly). By the year 2000 part of general fixing rubbers must be decreased to 69% and special rubbers part of must be increased to 24%. The production of latexes will increase from 5 to 7% [11]. In period of 1995-2000 the creation and the introduction in the action of new production capacities on 500 th.t./year are planned (Table 6). Assortment of new types of synthetic rubbers is shown in Table 7. It includes: halogenated (chloro and bromo) buthyl rubbers for the fabrication of nonchamber tyres, polyisoprene with cis-3,4-structure, butadiene rubbers, containing 1,2-unsaturated groups, elastomers instead of chloroprene which is produced by factory "Nairit" (Armenia), butadiene-nitrylic rubbers of new generation and others [24-26].

Table 5. The production of synthetic rubbers and latexes in Russia (th.t) [11]

Type of rubbers and latexes	Years			
	1991	1992	1995[x]	2000[x]
Rubbers of general fixing: isoprene SKI	831.0	675.9	560	620
butadiene SKD	316.1	241.9	220	250
butadiene-styrene SKS and SKMS	502.5	413.6	340	410
others	34.1	26.7	28.0	45
total	1683.7	1358.1	1148	1325
Rubbers of special fixing: butadiene-nytrilic SKN	84.4	55.1	60	70
buthyl rubbers BK	62.3	62.7	75	150
others	82.1	58.6	94	240
Latexes: SKS	84.3	64.1	73	123
others	15.5	12.2	15	26
total	99.8	76.3	88	149
on the whole	2012.3	1610.8	1465	1934

x - planned values

Table 6. The production capacities of synthetic rubbers and latexes [11]

Products	Factories	Capacity increase th.t/year
	New factories	
Buthyl rubber (BR)	JC "Tobolskyi OCC"	80
	"Sovbuthital" (Tobolsk)	60
Halogenated BR	the same	30
Ethylenepropylene Rubber "SKEPT"	IA "Nizhnekamsk-oil"	30
Thermoelastoplastics (TEP)	JC "Tobolskyi OCC"	60
	JC "Synthezcaoutchouk" (Toliyatty)	30
Divynilstyrene "DSSK"	the same	60
Polyisobuthylene	JC "Efremovskyi zavod SK"	15
Polyurethanes	JC "Chaikovskyi zavod SK"	25
Latex SKS	JC "SK-Premier" (Yaroslavl)	41
	IA "Kazanskyi zavod SK"	30
Latex DSVP	IA "Omskyi zavod SK"	7
Latex SKN	IA "SK-Premier" (Yaroslavl)	14
	Acting factories	
Buthyl rubber	JC "Synthezcaoutchouk" (Toliyatty)	from 35 to 43
Latex SKS	IA "Omskyi zavod SK"	from 9 to 12

note: JC is Joint Stock Company; OCC - Oil Chemical Combine; IA - Industrial Association

Table 7. The production perspectives of special fixing rubbers [11] (Thousand tons)

Rubber	Years	
	1995	**2000**
Ethylenepropylene "SKEPT"	4	24
Thermoelastoplastics	30	90
Chlorobuthyl rubber	–	20
Polyurethanes	1	11
Polyisobuthyelen	4	12
Chloroprene	–	20
SKOP	36	40

We have in Russian language some proverbs: "two 'perestroika' are proverbs equal one fire." We hope that are rearrangement time is going to end and we sure that in near future our industry will go up.

REFERENCES

1. A.A. Solovyanov, *Russian Chemical Journal*, 1994, vol. 38, N5, p. 3.
2. T.T. Rykuova, *Chemical Industry*, 1994, N4, p. 63 (in Russian).
3. *Financial news*, 1993, N17, pp. 18, 20 (in Russian).
4. M.V. Makarenko, V.L. Ulanov, *Chem. Industry*, 1993, N6, p. 59.
5. N.Ya. Usatchev, Kh.M. Minatchev, *Oil-Chemistry 1993*, vol. 33, N5, p. 387 (in Russian).
6. K.M. Dyumaev, *Chem. Industry 1992*, N12, p. 3.
7. G.S. Golovyn, *Russian Chem. J.*, 1994, vol. 38, N5, p. 7.
8. R.G. Melkonyan, *Chem. Industry 1994*, N6, p. 407.
9. A.I. Salyn, *Chem. Industry*, 1992, N5, p. 3.
10. V.P. Ivanov, *Chem. Industry*, 1993, N5, p. 3.
11. A.P. Nikandrov, et al. *Caoutchouk and Resin*, 1993, N5, p. 18 (in Russian).
12. V.L. Ulanov, M.V. Makarenko, *Chem. Industry 1994*, N1, p. 60.
13. E.A. Savvyna et al. *Plastmassy 1990*, N10, p. 3 (in Russian).
14. A.V. Polyakov, et al. *Plastmassy 1990*, N10, p. 11.
15. S.S. Stanovaya et al. *Plastmassy 1990*, N10, pp. 25, 28.
16. A.V. Shagilova et al. *Plastmassy 1990*, N10, pp. 31,32.
17. V.Ph. Vasylenko et al. *Plastmassy 1990*, N10, p. 35.
18. I.V. Peshkov, *Plastmassy 1991*, N4, p. 3.
19. B.B. Kudryavtzev *Varnish-Paint Materials and their Application* 1994, N6, p. 3 (in Russian).
20. *VPM and their Applications, 1993*, N2, p. 57; N6, p. 17.
21. Yu.A. Mikhailin, *Plastmassy 1993*, N3, p. 5.

22. R.A. Gommen et al. *Plastmassy 1990*, N8, p. 3.
23. E.A. Petrov, et al. ibid p. 16.
24. B.S. Grishyn *Caoutchouk and resin 1993*, N5, p. 3.
25. Yu.L. Morozov, ibid, p. 7.
26. G.A. Lysova, Yu.L. Morozov, *Caoutchouk and resin 1993*, N6, 7.

MATERIALS AND THE ENVIRONMENT

K.Z. Gumargalieva, G.E. Zaikov, Yu.V. Moiseev
Institute of Chemical Physics, Russian Academy of Sciences,
4 Kosygin Str., 117334 Moscow, Russia

"There are two possibilities - either mankind will make less smoke to be on Earth or the smoke will make less men to live." This phrase is very fair. The same idea Professor Jacques Yves Cousteau (France) has formulated in the following way: "Our globe resembles a car, running alone in the space with exhaust pipe revealed into passengers' saloon. The more force we apply to the accelerator, the higher the probability is to poison the driver and the passengers."[1] If we consider the factors, promoting degradation (aging) of polymeric materials, the most efficient ones among them are temperature (heat), air oxygen and light (especially ultraviolet) [1-6]. However, as the surroundings is polluted by the waste of human vital activities, polluted air and polluted water and soil begins to display more and more higher danger. In recent time only the attention was paid to decomposition of polymeric articles under the influence of the surroundings. If you pass along Malecon the main embankment of Habana City, it can be easily seen that houses are in poor conditions. Ferrum constructions became rusty-through, copper became green, paints peeled off. This is connected not only with poor state of economy, but with degradative influence of salty sea spray in presence of tropical climate with bright sunshine. Before scientists and skilled workers came to the idea, that pollutions in the atmosphere promote decomposition of polymeric articles, art representatives discovered the following regularity: metal and marble statues in Rome and Athens began to spoil with the increase of the amount of cars and pollution of the surroundings by their exhausts. It became necessary to place statues into museums, and they were changed by duplicates on their places (they are grudged not so serious).

---

[1] As it is known, the real system - the automobile with exhaust gases outlet into passengers' compartment, was created by Hitler's SS service for killing prisoners of pens and concentration camps.

In the present paper we present some experimental data about the influence of the surroundings (pollutions of the surroundings, to be more correct) on the accelerated aging of metal and polymeric materials. In this case special acsent will be devoted to the experimental data, obtained in Russia. The aim of this article is to attract attention of scientists and skilled workers once more to the necessity of investigations of corrosion processes in polymers and metals under the influence of the surroundings, to attract special attention to the study of kinetics and the mechanism of degradation processes and searching the ways of stabilization.

Table 1 represents the structures and mistakes of estimation of exhausts of SO_2 and nitrogen oxides per year on the total square of Russia [7]. Figures 1A and 1B exhibit European maps for 1984 and 1994 and fall-outs occurring with SO_2 (heat electric power stations and automobile exhausts mainly). As it is seen from the Figures a definite stability of this problem is observed in Western Europe. At the same time air pollution by sulfur oxides (as well as by nitrogen ones) increases in Russia and Ukraine. First of all this fact is connected with growing of the amount of cars (mainly old and the ones without catalysts of afterburning at the exhaust). The paradox of Russia is in the fact that the most people live worse every year, and the amount of cars of Russian new comrades ("New Russians") becomes greater.

Table 1. Sources of SO_2 and NO_2 exhaust a year over the territory of Russia [7] in 1990

Source	Exhaust, % SO_2	error, % SO_2	Exhaust, % NO_2	error, % NO_2
Electric power stations	55	15	20	20
Heating systems	20	15	20	20
Industrial ventures	15	20	5	20
Transport	10	30	55	30
All	100		100	

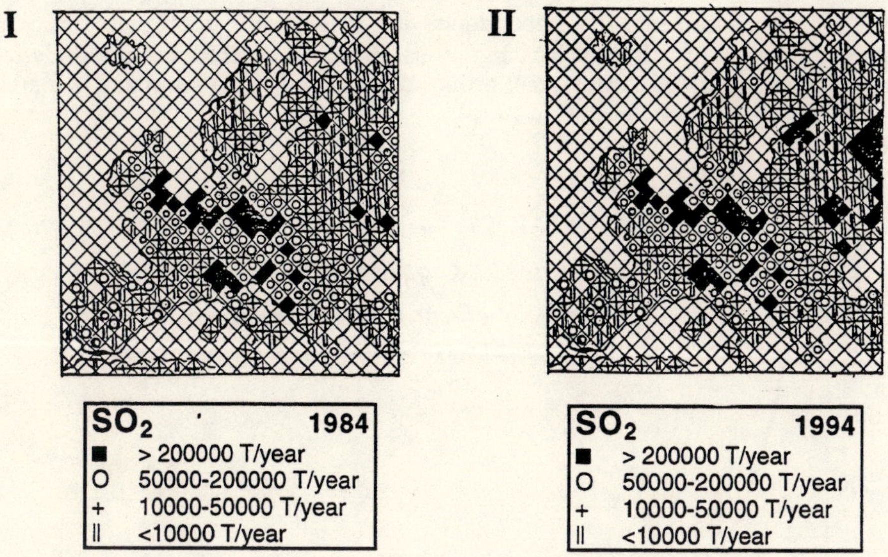

Figure 1. SO_2 exhaust in atmosphere (tonn/km^2 per year) I-1984, II-1994, ■>200000 tonn/year, o-50000-200000 tonn/year, +10,000-50,000 tonn/year, | | <10,000 tonn/year.

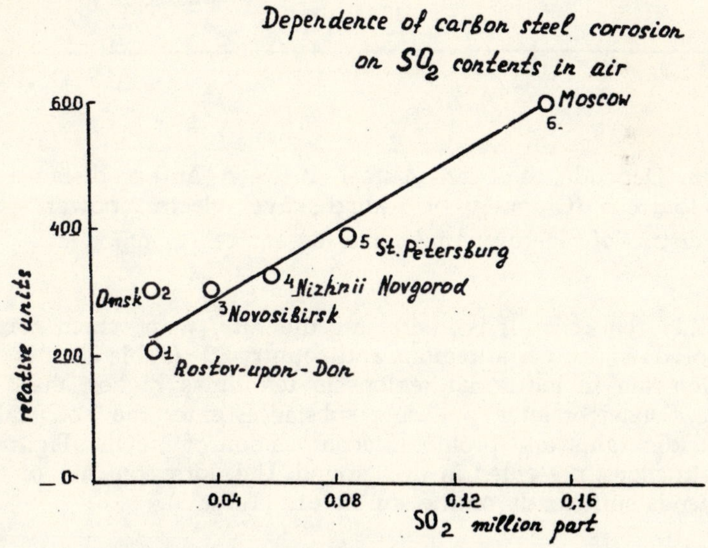

Figure 2. Dependence of carbon steel corrosion (relative units) on SO_2 contents in air. 1 - Rostov upon Don, 2 - Omsk, 3 - Novosibirsk, 4 - Nizhnii Novgorod (Gorkii), 5 - St. Petersburg, 6 - Moscow.

Figure 2 presents the dependence of corrosion of carbon steel on SO_2 content in the air of different Russian cities. Figure 3 exhibits the dependence of the corrosion rate of carbon steel on the distance from the source of exhaust of polluting gases (power station pipe).

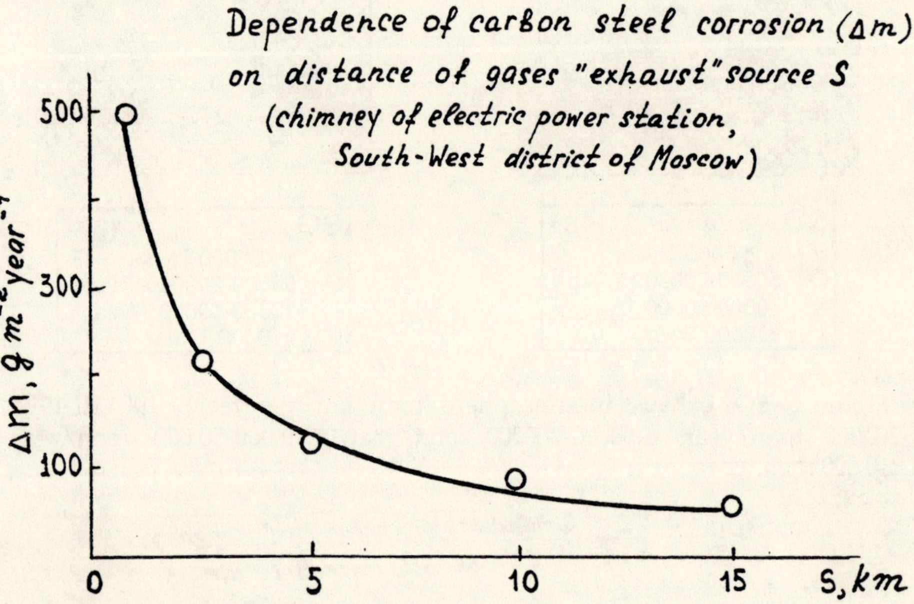

Figure 3. Dependence of carbon steel corrosion (Δm) on distance of gases "exhaust" source S (Chimney of Kaluzhskaya electric power station at Southwest district of Moscow) Δm - corrosion of steel (g/m^2).

In this connection it is interesting the rate, with which corrosion of metals proceeds in industrial regions and countryside (Table 2). It is seen that the corrosion rate in industrial regions is ten times higher than that in countryside. Sooner or later polluting substances enter the soil, lakes and rivers as acidic rains and promote decomposition of metals (Figure 4) and metal constructions, presented in the ground. The corrosion rate of different metals depends sufficiently on the soil nature (Table 3).

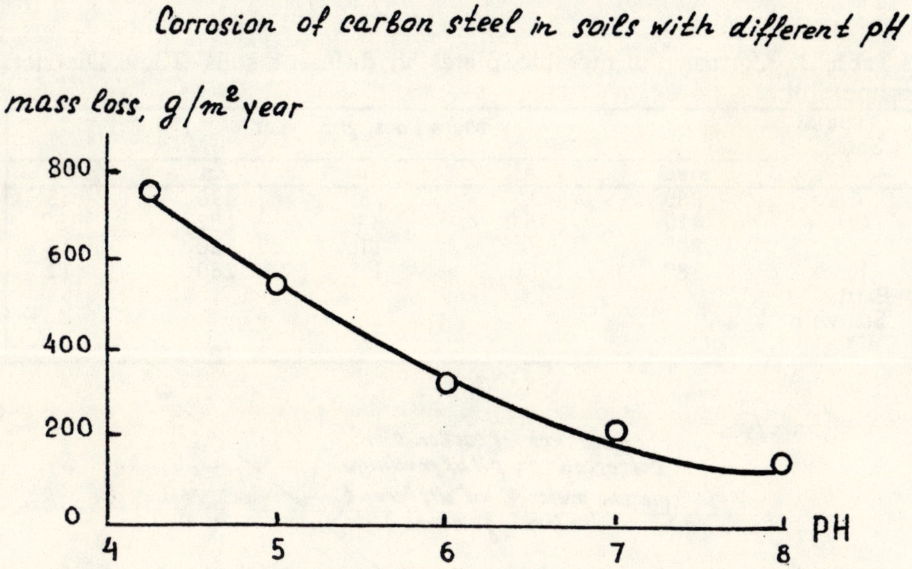

Figure 4. Corrosion of carbon steel in soil with different pH (fields of Moscow region).

Table 2. Rate of carbon steel corrosion at different testing stations of Europe [7], [8]

Conditions	Location of Testing Station	Rate of Corrosion g/m^2 year
Industrial	Most (Czech Rep.) Anby (France) Kopist (Czech Rep.) Duisburg (Germany) Lugansk (Ukraine) Novokuznetsk (Russia)	275 350 365 490 510 555
City	Prague Stockholm Moscow	196 210 240
Country	Picherande (France) Belaya tserkov' (Ukraine) Flahult (Sweden) Zvenigorod (Russia) Kasp. Horey (Czech Rep.) St. Denis (France)	40 50 55 60 65 125

Table 3. Corrosion of metallic plates in different soils (Kiev District, Ukraine), [8]

Type of Soil	Mass Loss, g/m^2 year				
	steel	Al	Cu	Zn	Pb
Clay	360	3.7	25	490	45
Clay + Sand	410	15.0	33	102	150
Black Earth	305	0	11	120	15
Soil with $CaCO_3$	180	1.0	11	280	11

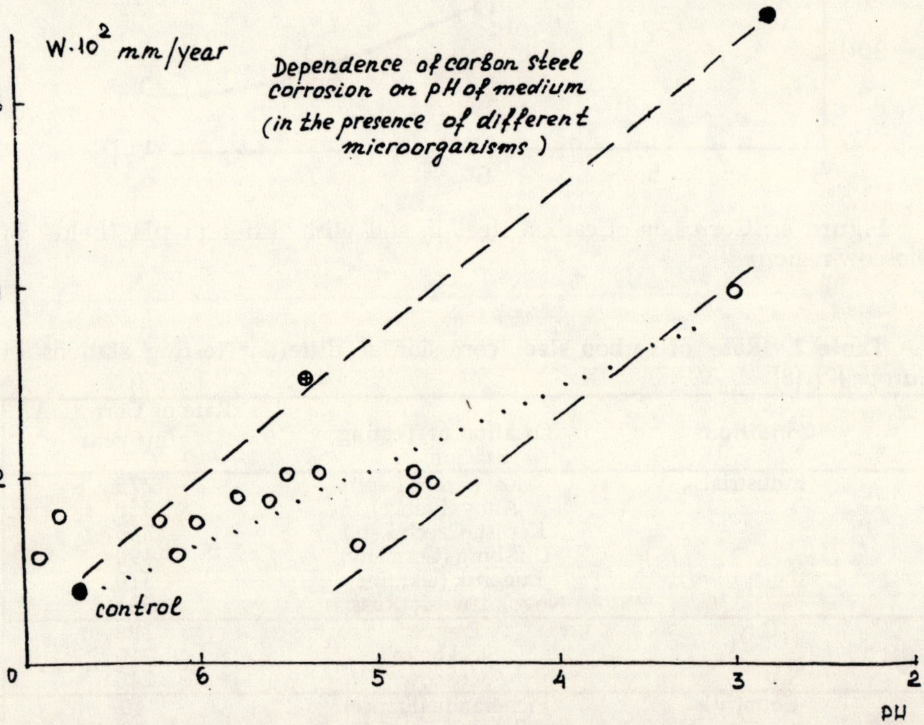

Figure 5. Dependence of carbon steel (Russian standard No. 3) corrosion on pH of medium (pH of medium is created by microorganisms).

Considering soils, we should not forget about microorganisms, which themselves may create a definite pH of the medium as a result of their vital activities (Figure 5). Many bacteria (see for example Figure 6) do often accelerate corrosion. The role of sulphate-reducing bacteria is the most sufficient, because they sufficiently accelerate the corrosion process (Figure 7). All said about metals concerns sufficiently polymers also (plastics, caoutchoucs, rubbers, fibers, polymeric covers, etc.). As an example Figure 8 presents the influence of exhaust gases from Russian "Volga" car on the rate of "Nylon 6" degradation. It is seen that the rate of degradation (which was controlled by tensile strength) is sufficiently higher in exhaust atmosphere, than under the influence of air oxygen or UV-rays. In this case accelerating influence on the process of Nylon 6 degradation is performed by both SO_2 and nitrogen oxides (Figure 9).

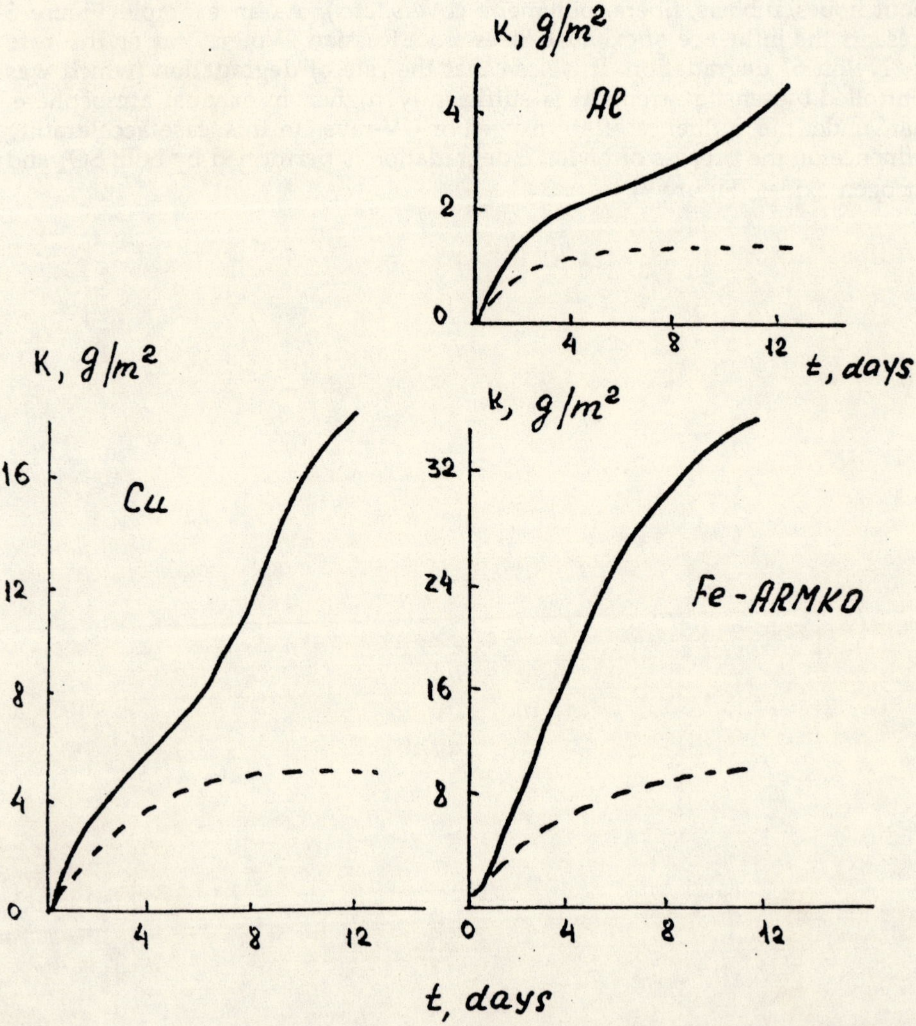

Figure 6. Influence of fungi Aspergillus niger on corrosion of Al-ABI, Cu-MB and FC-ARMKO. Dotted line shows control experience (without Aspergillus niger).

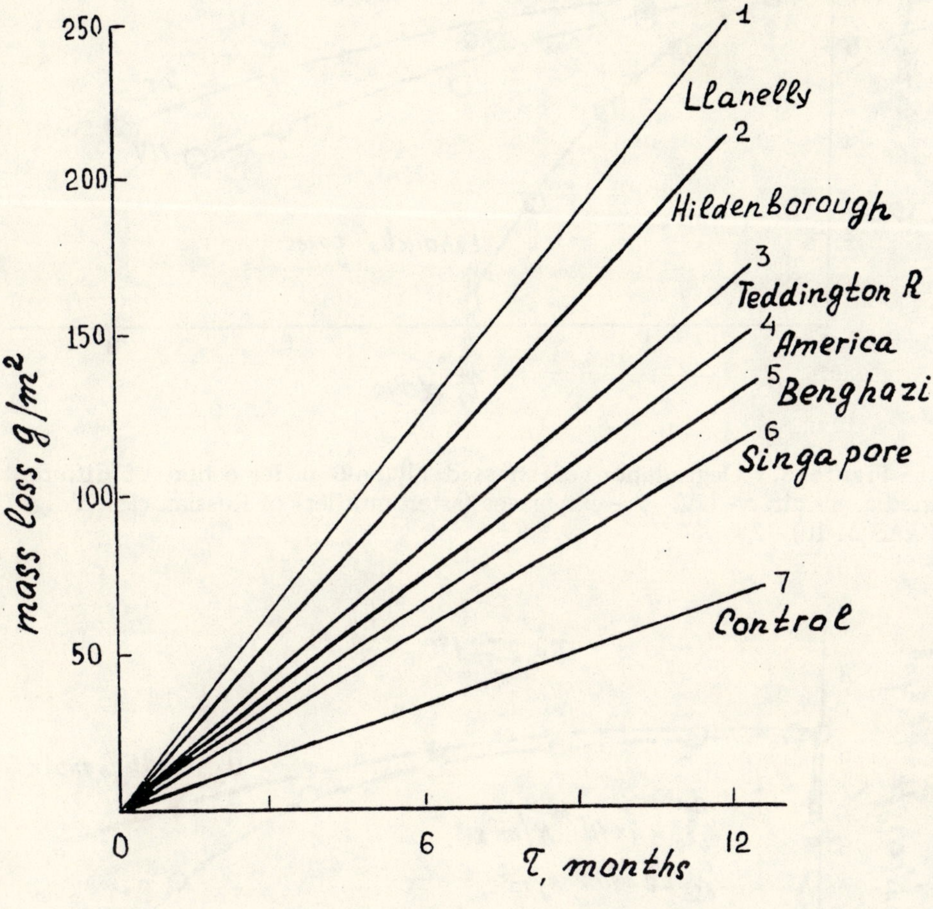

Figure 7. Corrosion of steel plate under action of different sulphate reproduce bacteria.

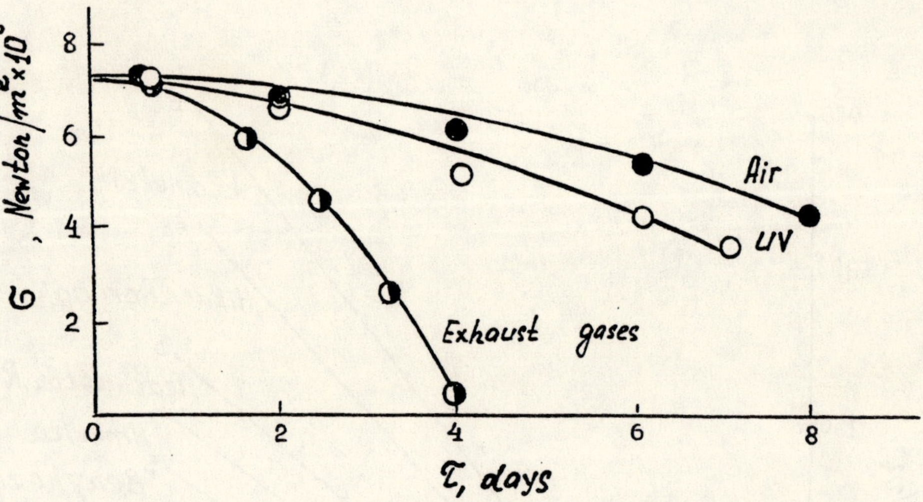

Figure 8. Degradation of stressed Nylon-6 under action of different media: • - air, o - UV, - exit gases (after muffler) of Russian car "Volga" (GAS-24-10).

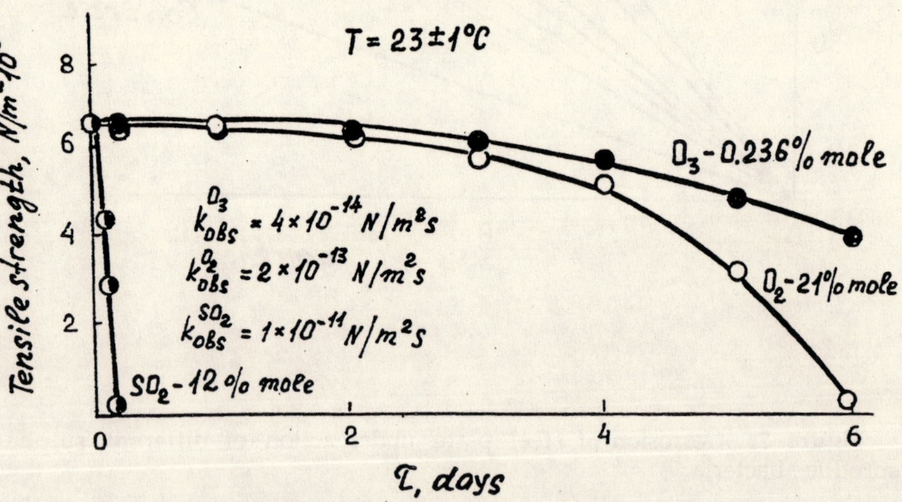

Figure 9. Degradation of Nylon-6 in different gaseous media: O_3-0.236 % mole - •, O_2-21% vol-o, SO_2-12.00% vol- T=23±1°C, $k_o{}^3{}_{obs}$=4x10^{-4}.

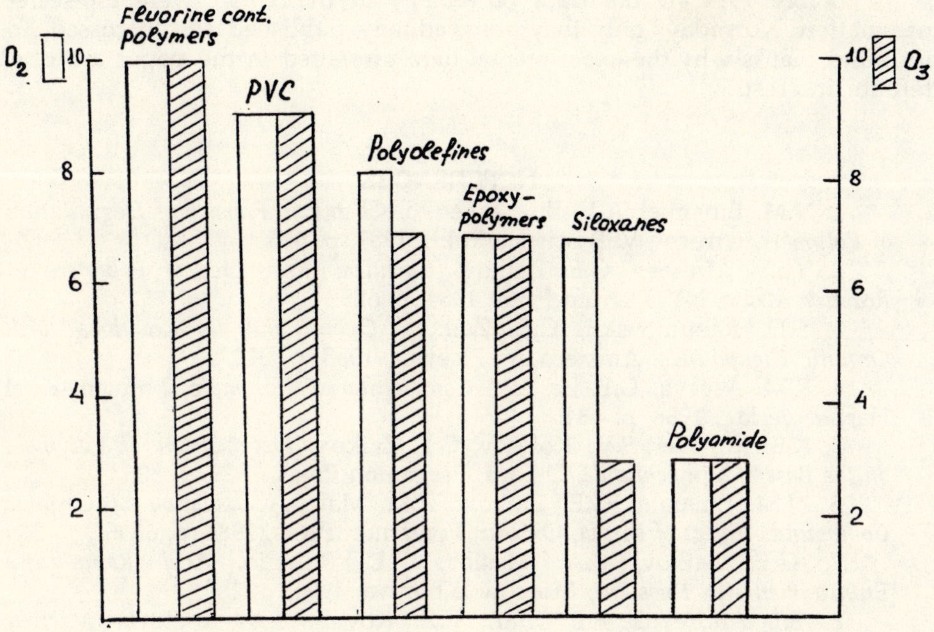

Figure 10. Coefficient of stability for different polymers in the presence of O_2 and O_3. ❑ - O_2, ■ - O_3.

Figure 10 exhibits coefficients of stability of different polymers in acidic media [2,3] in relation to the influence of oxidizers (O_2 and O_3). There are shown no data on ozone aging of caoutchoucs and rubbers. Rate constants of ozone reaction with double C=C bonds is 1000 - 10000 times higher than that with single C-C bonds [3]. That is why ozone is very dangerous for caoutchoucs and rubber as the product of pollution of the surroundings (photochemical smog) and requires highly effective antiozonants, able to suppress its influence.

It should be reminded, that heterochain polymers are the most vulnerable from the point of view of acidic rains (polyamides, polyesters, polyacetales, etc.), which are easily hydrolyzed in acidic medium [2]. Considering nitrogen oxides, polymers with double bonds are sufficiently instable to their influence as well as the ones inclined to cross-linking. Among these polymers we can name polymethylmetacrylates and their derivatives.

Before 1991 all the data on ecology in the USSR were top-secret information. Nowadays only they are gradually published and discussed. To our mind that is why the experimental data presented in the paper are of a definite interest.

REFERENCES

1. N.M. Emanuel, A.L. Buchachenko, *Chemical Physics of Degradation of Polymers*, Utrecht, VSP Science Publ., 1987, p. 425.

2. Yu.V. Moiseev, G.E. Zaikov, *Chemical Resistance of Polymers in Reactive Media*, NY, Plenum Press, 1984, p. 625.

3. S.D. Razumovskii, G.E. Zaikov, *Ozone and its Reactions with Organic Compounds*, Amsterdam, Elsevier, 1985, p. 392.

4. R.M. Aseeva, G.E. Zaikov, *Combustion of Polymers*, Munich, Karl Hanser verlag, 1988, p. 434.

5. K.S. Minsker, S.V. Kolesov, G.E. Zaikov, *Degradation of Polymers on the Base of Vinylchloride*, Oxford, Pergamon Press, 1989, p. 526.

6. N.M. Emanuel, G.E. Zaikov, Z.K. Maizus, *Oxidation of Organic Compounds. Effect of Media*, Oxford, Pergamon Press, 1984, p. 6646.

7. G.E. Zaikov, S.A. Maslov, V.L. Rubailo, *Acid Rains and Environment (in Russian)*, Moscow, Khimiya, 1991, p. 159.

8. E.I. Andreyuk, P.I. Bilai, E.Z. Koval', I.A. Kozlova, *Microbe corrosion (in Russian)*, Kiev, Naukova dumka, 1980, 362 p.

APPLICATION OF THE MIMIMAX METHOD FOR CALCULATION OF VISCOELASTIC MATERIAL FUNCTIONS AND RELAXATION SPECTRA OF POLYMER MELTS AND SOLUTIONS

Yu.G. Yanovsky and Yu.A. Basistov
Institute of Applied Mechanics of Russian Academy of Sciences
Leninsky prospect 32A, Moscow 117334, Russia

Abstract — Many publications were devoted to the problem of finding methods for calculating the distribution function of relaxation time spectra H(τ) for viscoelastic media. Direct calculations of this function on the basis of known relations of the linear theory of viscoelasticity connected with the solution of Fredholm integral equation of the first kind involve serious difficulties. At the same time, numerical approaches to determining the function H(τ) based on the use of experimentally determined material functions were proposed in a number of recent publications. Since the accuracy of calculating the functions depends directly on the confidence of determining the primary experimental data that generally always contain errors of a statistical nature, a number of authors employ smoothing numerical methods of determining the material functions, or ones that fail to react to the appearance of rough errors. These methods include methods of regularization and the maximum of entropy. However, they have a number of serious shortcomings limiting the possibilities of their successful use.

In the present work, we propose to employ a different numerical method for these purposes, namely, the minimax technique. In our opinion, it is more accurate, rapid, and universal. We consider a scheme of constructing an algorithm, give examples of solving test problems showing how the algorithm functions in various extremal hypothetic situations. The results of the numerical calculations of the function H(τ) for a real object, viz. polybutadiene are based on experiments run in a rotary viscometer Rheotron of the firm Brabender.

INTRODUCTION

The viscoelastic reaction (behavior) of a system to deformation is a widely known experimental fact. From a physical viewpoint, such behavior is generally explained either by the existence of an internal structure, or by the heterogeneous structure of the material, or by both simultaneously. It features the most diverse systems, namely, polymers (at temperatures above the glass transition one), composites based on polymers, hybrid composites, fluid systems of the type of highly structurized petroleums, suspensions, etc.

Unlike Hookian bodies or Newtonian liquids, the problem of describing the properties of such systems within a broad range of variation of the magnitudes and laws of deformation, the rates (or frequencies) of deformation, and other parameters generally consists in determining the time dependences of the mechanical characteristics, i.e. the material functions reflecting the viscoelastic mechanism of the behavior of these systems. In terms of moduli and viscosities and within the framework of a linear approximation (conditions not depending on the deformation amplitude), these characteristics can include the relaxation modulus G(t),where t is the time, or the dynamic storage $G'(\omega)$ and loss $G''(\omega)$ moduli, where ω is the angular frequency of deformation. The storage and loss moduli are the real and imaginary components of the complex dynamic modules $G^*=G'+iG''$.

The time dependences of G can be obtained quite simply from experiments involving stress relaxation wherein the dynamics of the change in the modulus is studied after the action of deformation is stopped. The values of $G'(\omega)$ an $G''(\omega)$ are determined for conditions of the periodic action of deformation set by the law $\gamma=\gamma_0 e^{i\omega t}$, where γ_0 is the deformation amplitude.

From the standpoints of a model approach, the ability of a viscoelastic body to resist a mechanical load and react thereto adequately is determined by a very important characteristic of a system, viz, the relaxation time τ_i. As indicated in Reference [1], the relaxation time is determined by the ratio of the elastic and viscous (dissipative) properties of a material and in the simplest case can be introduced as

$$\tau_i=\eta_i/G_i$$

where η_i is the viscous and G_i the elastic components.

Such a model approach is simplified for real viscoelastic media that generally have a rather complicated internal structure. It is quite natural to introduce a distribution function of the relaxation time, viz. the relaxation spectrum $H(\tau)$. Experiments revealed [1,2] that for a broad class of materials (e.g. polymer solutions and melts, polymer blends, polymer-based composites), the range in the variation of the relaxation time may form many decimal orders of magnitude. The mechanical reaction of viscoelastic media to deformation is determined exactly by the features of this spectrum. Depending on the parameters of the external action (the laws of deformation, the rates or frequencies of deformation, temperatures, etc.), these media may reveal the properties of fluid, rubbery, leathery, or glassy bodies.

Establishment of the form of the function $H(\tau)$ is an important theoretical and practical task, in the former case for constructing physically substantiated theories (phenomenological or molecular) of the viscoelastic behavior of such systems, and in the latter for predicting the properties of specific materials.

By the linear theory of viscoelasticity, the material functions G(t), G'(ω), and G"(ω), and also the relaxation spectrum of the system H(τ), where τ is the relaxation time, are related by the following expressions [1]:

$$G(t) = G_e + \int_{-\infty}^{\infty} H(\tau) e^{-t/\tau} d\ell n\tau, t\varepsilon(-\infty,\infty); \quad (1)$$

$$G'(\omega) = G_e + \int_{-\infty}^{\infty} H(\tau) e^{-t/\tau} d\ell n\tau, t\varepsilon(-\infty,\infty); \quad (2)$$

$$G''(\omega) = G_e + \int_{-\infty}^{\infty} [\omega^2\tau^2/(1+\omega^2\tau^2)] H(\tau) d\ell n\tau, \omega t(-\infty,\infty); \quad (3)$$

$$G''(\omega) - iG'(\omega) = \omega \int_{-\infty}^{\infty} G(t) e^{-i\omega t} dt, \omega\varepsilon(-\infty,\infty). \quad (4)$$

The constant G_e added in expressions (1) and (2) considers the contribution of the discrete component into the spectrum for viscoelastic solids at τ→∞. For viscoelastic liquids and uncrosslinked polymers above the glass transition temperature Ge=0.

It can be seen from expressions (1)-(4) that from the theoretical viewpoint the determination of at least one of the functions G(t), G'(ω), G"(ω), and H(τ) indicated above allows one to calculate the others. In practice, however, these calculations associated with integration of expressions (1)-(4) involve considerable mathematical difficulties, and consequently the obtaining of an exact solution is problematic.

From an experimental viewpoint, the possibility of determining the functions G(t), G'(ω), and G"(ω) within quite broad ranges of t or ω depends on the features of the instruments employed. Quite often we succeed in establishing the form of a material function only within relatively narrow ranges of times or frequencies that do not overlap mutually. (The restrictions generally concern either relatively high values of t (or low ones of ω), or relatively low t's (or high ω's)).

Having in view that direct experimental appraisals of the relaxation spectrum H(τ) are impossible and it can be calculated only from material functions (see expressions (1)-(4), and also References [1-3]), and in the latter case the use of the results of various independent experiments is tempting, the development of numerical methods for such calculations is a very urgent task.

Since the direct calculation of the function H(τ) is associated with the confidence of the experimental determination of the relations G(t), G'(ω) and G"(ω), we shall briefly analyze below the possible experimental errors in determining the material functions we are interested in.

EXPERIMENTAL ERRORS AND APPROACHES TO THEIR SIMULATION

Experimental results always contain errors, chiefly of a statistical nature. They are explained by the influence of parameters not controlled during measurements. It is exactly the errors of a statistical nature that can tell considerably on the authenticity and correctness of interpretation of the results. Statistical errors are usually simulated by independent random quantities from the Gaussian distribution:

$$f(x)=\frac{1}{\sigma\sqrt{2\pi}}\exp\left\{-\frac{(x-M)^2}{2\sigma^2}\right\},\ x\varepsilon(-\infty,\infty),$$

where δ^2 and m, the dispersion and mean, are considered to be known *a priori*.

There are several tests for verifying hypotheses on a normal (Gaussian) distribution [4-6]. One of them is the Kolmogorov-Smirnov test (see, for example, Reference [5]). Another one, the W-test, was introduced by Shapiro and Wilk and was used in Ref. [4] and [6]. In the latter, the W-test was calculated by the expression

$$W=\frac{\left[\sum_{i=1}^{n}a_i\eta(i)\right]^2}{\sum_{i=1}^{n}[\eta(i)-\eta]^2}, \tag{6}$$

where $\overline{\eta}=\frac{1}{n}\sum_{i=1}^{n}\eta(i)$ is the sample mean,

$\eta(1)\leq\eta(2)\leq\ldots\leq\eta(n)$ is the ordered sample,

$a=(a_1,\ldots,a_n)^T=m^TV^{-1}[(m^TV^{-1})(V^{-1}m)]^{-\frac{1}{2}}$ is the vector of the parameters in (6), T is transposition, m - (n x l) is the mean vector, and V - (n x n) is the correlation matrix of the n-dimensional normal distribution.

However, when the tests indicated in Ref. [4-6] are employed, there is also *a priori* indefiniteness with respect to the dispersion δ and mean m of this normal (Gaussian) distribution (see Equation (5)), which always exists. This hinders the compilation of an algorithm of a process and its correct numerical realization. Moreover, the postulate itself on the correctness of the law of Gaussian distribution of errors was also considered to be doubtful in a number of publications [7].

If we presume that errors of a statistical nature always include freak values, the W-test in a number of cases may be unsuitable. Indeed, the appearance of a freak value in statistics $\{\eta(i)\}_{i=1}^{n}$ mainly determines the result of summation when calculating the sample mean $\bar{\eta}$. Moreover, indefiniteness remains with respect to m and V. In this case, to obtain confident results according to experimental data, one must employ smoothing or a method that does not react to the appearance of rough errors. These methods include first of all the method of regularization [8,9] and that of the maximum of entropy [10].

It should be noted that the methods indicated in [8-10] are known to be ones for solving problems incorrect according to Hadamard. The solution xεX with respect to the data of yεY is considered to be correct according to Hadamard if for any yεY the solutions x=f(y)εX: a - exists, b - is the only one, and c - is stable, i.e. $\rho_x(\tilde{x},\bar{x})\to 0$ when $\rho_y(\tilde{y},\bar{y})\to 0$. Here $\rho_x(°)$ and $\rho_y(°)$ are valuations of the spaces X and Y, respectively, while $\tilde{x},\tilde{y}$, and $\bar{x},\bar{y}$ are approximate and unknown accurate values, respectively. Violation of one of the three conditions indicated above makes the problem incorrect.

Without refuting the usefulness of the methods of regularization and maximum of entropy employed, in particular, in Ref. [4,6,10] to calculate the relaxation spectra of selected viscoelastic media, we shall note that they have a number of serious shortcomings (they will be treated below) as a result of which their use does not always yield correct results.

As applied to the problems of the present work, it would be better, in our opinion, to use a different method of solving incorrectly formulated problems, namely, the minimax one [11]. Unlike the methods in Ref. [6,8-10], the minimax method yields the best (optimal) solution with the least favorable initial data. We shall consider further the features of employing this method for calculating the function H(τ) by the values of material functions determined from rheological experiments.

MINIMAX METHOD, BRIEF DESCRIPTION OF ALGORITHM

From the standpoints of a numerical approach, the problem consists in calculating a function, e.g. H(τ), from experimental data containing outliers (freak values). We shall select the set of possible values of the loss modulus $\tilde{G}_i^{"}$ as these data. The values were obtained in the range of frequencies ω_i and contained the outliers $\xi(\omega_i)$, i.e.

$\tilde{G}_i^{"} \equiv \overline{G}"(\omega_i) + \xi(\omega_i)$, when i=l, ..., N

When the outliers $\xi(\omega_i)$, i=l,...,N, are present, the problem of evaluating $H(\tau)$ is a priori indefinite with respect to the given $G"(\omega_i)$ and, consequently, requires additional *a priori* information. The latter can be introduced by the value set

$$\left\{\tilde{G}_i^{"}\right\} : \frac{1}{N}\sum_{i=1}^{N}(\overline{G}_i^{"} - \tilde{G}_i^{"})^2 \leq \delta^2, \tag{7}$$

where $\left\{\overline{G}^{"}\right\}_{i=1}^{N}$ are the unknown exact data, and $\delta \geq 0$ is a parameter of the error of the data.

With a view to the restrictions (7), the optimization problem acquires the form:

$$H^{\delta}(\tau) \approx \arg\min_{H} \max_{\tilde{G}"}(1/N)\sum_{i=1}^{N}\left[G_i^{"} - \int_{-\infty}^{\infty}\frac{\omega_i\tau}{1+(\omega_i\tau)^2}H(\tau)d\tau\right]^2 \tag{8}$$

The use of (8) with the restrictions (7) guarantees the calculation of the approximation $H^{\delta}(\tau)$ converging to the unknown accurate solution $H(\tau)$ in a root-mean-square approach when $\delta \to +0$.

It should be noted that with the corresponding selection of the positive-negative Hilbert spaces, this convergence can be guaranteed in a uniform valuation with a certain degree of smoothness [11].

The error parameter $\delta>0$ in problem (7), (8) as in the method of regularization, is considered to be known *a priori*. Nevertheless, as in the methods of regularization and maximum entropy [8-10] we may also obtain here a quite "poor" solution $H^{\delta}(\tau)$ if the values of the parameters $\delta>0$ are set *a priori* by perturbations of the initial data not corresponding to the real values. This is why the error parameter $\delta>0$ must be appraised by experimental results when employing methods of solving incorrectly posed problems.

For this purpose, let us consider a sequence of random values:

$$\tau_i = \tilde{G}_i^{"} - \int_{-\infty}^{\infty}\frac{\omega_i\tau}{1+(\omega_i\tau)^2}H^{\delta}(\tau)d\tau, i=1,\ldots,N \tag{9}$$

where $H^{\delta}(\tau)$ is the solution of problem (7), (8) for a value of δ chosen arbitrarily beforehand. We shall adopt the following assumptions relative to the random values of r_i, $i=1,\ldots,N$: the values $\{r_i\}$ are statistically independent and satisfy an ε-class of distribution functions [12]:

$$\phi=\left\{F_{\varepsilon}(x)=(1-\varepsilon)F(x)+\varepsilon P(x),\,\varepsilon\tau[0,1]\right\}, \tag{10}$$

where F(•) is the theoretically set distribution function, and P(•) is an unknown distribution function considering the presence of freak values.

The numerical parameter $\varepsilon\in[0,1]$ can be interpreted as the probability of appearance of freak values in the experimental results. When $\varepsilon=0$, we have the case of indefiniteness with respect to the parameters F(x) or complete a priori definiteness (see, for example, Ref. [6]). When $\varepsilon=1$, we have the case of nonparametric a priori indefiniteness wherein the distribution function P(x) cannot be determined a priori. And, finally, the case $0<\varepsilon<1$ takes into account the parametrically-nonparametrically a priori indefiniteness.

Let us assume that the ε-class of distribution functions (10) has symmetric densities $f_{\varepsilon}(x)=f_{\varepsilon}(-x)$ with Fisher's

$$I(f_x)=\int_{-\infty}^{\infty}(f_{\varepsilon}'/f_{\varepsilon})^2 dF_x(x)<\infty,\,\varepsilon\in[0,1],\,I(f_x)<\infty$$

We consider the density f(x) to be normal

$$f(x)=\frac{1}{\delta\sqrt{2\pi}}\exp\left\{-\frac{(x-m)^2}{2\delta^2},x\in(-\infty,\infty)\right\}$$

with an a priori unknown mean m and dispersion δ^2. The latter characterizes the parametric indefiniteness of the problem and is well consistent with practice. The density p(x), $x\in(-\infty,\infty)$ is considered to be symmetric and unknown and takes into account the appearance of freak values for both large and small values of X (the requirement of symmetry). The latter reflects the nonparametric a priori indefiniteness of the problem. The quantity $\varepsilon\in[0,1]$ in (10) determines the fraction of freak values in an infinite volume of data.

Consequently, it is necessary to appraise the parameters (m,δ) of a hypothetic normal distribution according to a finite sample of a fixed volume with parametric-nonparametric indefiniteness.

Let us introduce the minimax variation problem:

$$\begin{cases} (L^*, F^*) = \arg \min\limits_{L \in G} \max\limits_{F \in \phi} V^2(L, F), \\ V^2(L,F) = \left(\int\limits_{-\infty}^{\infty} \ell^2(x) dF) / [C(\int\limits_{-\infty}^{\infty} \ell_s^{'}(x-s) |_{s=0} \, dF \right)^2 \end{cases} \quad (12)$$

for any L∈G and F∈Φ. Here G is the set of convex functions L(t), t∈(-∞,∞) symmetric relative to the origin of coordinates and with continuous derivatives l(t)=dL(t)/dt, t∈(-∞,∞) and such that for any F∈Φ the inequalities

$$\int\limits_{-\infty}^{\infty} \ell^2(x) dF < \infty, (d/ds) \int\limits_{-\infty}^{\infty} \ell(x-s) dF < \infty$$

are observed for any Sε(-∞,∞).

Without delving into the mathematical details, with which one can acquaint oneself in Ref. [13], we shall give the final algorithm for appraising the parameters m and δ.

$$\begin{cases} \hat{m} = \dfrac{1}{N} \sum\limits_{i=1}^{N} \varphi(r_i), \\ \hat{\delta}^2 = \dfrac{1}{N} \sum\limits_{i=1}^{N} K(\dot{r}_i), \end{cases} \quad (13)$$

where r_i=r_i-m are centered random values.

The nonlinear without memory transforms φ(∞) and χ(∞) minimize Fisher's asymptotic information for the distribution function that is the least favorable in the sense of (12) and has the following form:

$$\varphi(x) = \begin{cases} \hat{S} - Km\hat{\delta}, x < \hat{S} - Km\hat{\delta}, \\ x, \quad \hat{S} - Km\hat{\delta} \le x \le \hat{S} + Km\hat{\delta}, \\ S + Km\hat{\delta}, X > \hat{S} + Km\hat{\delta}, \end{cases} \quad (14)$$

where $\hat{s} = med\{x_i\}_{i=1}^{N}$ is a sample median, and $\delta = med\{|x_i - \hat{s}|\}_{i=1}^{N}$ is the absolute median deviation.

Accordingly:

$$x(x)=\begin{cases}(\hat{\delta}\,\mathrm{Re}\sqrt{1-K_\delta})^2-1, & |x|<\delta(\mathrm{Re}\sqrt{1-K_\delta})\\ x^2-1 & ,\hat{\delta}(\mathrm{Re}\sqrt{1-K_\delta})\le|x|\le\hat{\delta}\,|\sqrt{1+K_\delta})\\ (\hat{\delta}\,|\sqrt{1+K_\delta}\,|)^2-1, & |x|>\hat{\delta}\,|\sqrt{1+K_\delta}\,|\end{cases} \quad (15)$$

The parameters K_m and K_o are consistent with the indefiniteness parameter e with the aid of the relevant coupling equations:

$$\begin{aligned}&\int_{-Km}^{Km}\exp\{-x^2/2\}dx+(2/Km)\exp\{-K^2/2\}=\sqrt{2\pi}/(1-\varepsilon)\\ &\int_{\mathrm{Re}\sqrt{1-K\delta}}^{\sqrt{1+K\delta}}\exp\{-x^2/2\}dx+[\mathrm{Re}\sqrt{1-K\delta}\exp\{-0,5\,\mathrm{Re}\sqrt{1-K\delta}+\\ &+|\sqrt{1+K\delta}\,|\exp\{-0,5\,|\sqrt{1+K\delta}\,|\}]/K\delta=\sqrt{2\pi}/[2(1-\varepsilon)]\end{aligned} \quad (16)$$

Equations (16) have single solutions at $[0,\infty]$ for any $\varepsilon\in[0,1]$. It should be noted that when $K_m\to\infty$ and $K_\delta\to\infty$ the algorithm (12)-(16) coincides with the well known algorithm of maximum probability. When $K_m\to+0$ and $K_\delta\to+0$, the algorithm (12)-(16) becomes nonparametric, the median remaining as an estimate of the mean, and the absolute median deviation as an estimate of the dispersion.

Numerical realization of the algorithm of the minimax problem when solving integral equations of the first kind.

As noted above, our task consists in obtaining an exact numerical solution of integral equations of the first kind of the type of (1-4), or in a more general form:

$$u(t)=\int_a^b k(t,\tau)z(\tau);\; t\in[c,d]; \quad (17)$$

with approximately given initial data:

$\tilde{K}[t,\tau],\ \tau\in[a,b],\ t\in[c,d],\ \tilde{u}(t),\ t\in[c,d]$ such that

$$\int_a^b \int_c^d |\tilde{K}(t,\tau) - \overline{K}(t,\tau)| \, dt d\tau \le h^2,$$

$$\int_c^d |\tilde{u}(t) - \overline{u}(t)|^2 \, dt \le \delta^2$$

We shall seek the exact solution of Equation (17) in the form:

$$z(t) = (1 - t^2)^2 + 0{,}5(\sin(\pi t))^4 + \cos(6{,}5\pi t),\ t \in [-1,1] \tag{18}$$

Analysis of function (18) reveals that it contains seven maxima and six minima in the region of determination. This allows us to analyze quite reliably each of the solutions obtained in uniform valuation in a series of consecutive approximations to the exact solution. We shall select the following function as the exact description of the nucleus in Equation (17):

$$K(t,\tau) = \frac{1}{1 + 100(t-\tau)^2},\ t \in [-1,1],\ \tau \in [-1,1]. \tag{19}$$

Function (19) is characterized by the fact that the matrix $\{K(t_i, \tau_i)\}^{M,N}$ corresponding to it has an inverse one, but which is poorly stipulated, because $\lambda_{max}/\lambda_{min}>>1$. Here λ_{max} and λ_{min} are its maximum and minimum eigenvalues, respectively.

An exact function of $\overline{u}(t)$ (Equation (17)) can be obtained by integrating it with a view to (18) and (19). To simplify this task, let us assume that h=0, the parameter $\delta>0$ is unknown, while the approximate data are determined by the relation

$$\overline{u}(t) = u(t) + \xi(t),\ t \in [-1,1]. \tag{20}$$

Here $\xi(t) \approx \{\xi(t_i)\}_{i=1}^{N}$ is a random function (an outlier in the experimental results) that is simulated by independent random numbers from two normals distributions with zero means and unknown dispersions by the relation:

$$\xi(t) = (1-\varphi)\xi_1(t) + \varphi\xi_2(t),\ t \in [-1,1] \tag{21}$$

In expression (21), is a random number equal to 0 or 1, while $\xi_1(t)$ and $\xi_2(t)$ simulate the random noise and their outliers in the experimental results or due to interruptions in the operation of the instruments, respectively.

Expression (21) can be seen to be identical to (10) at a certain preset value of $\varepsilon \in [0,1]$. With a view to (8), the task of optimization for Equation (17) will be written as follows:

$$z^{\delta}(t) = \arg \min_{z \in w_2^1} \max_{u \in L_2} \left\{ \int_c^d \left[\tilde{u}(t) - \int_a^b K(t,\tau) z(\tau) d\tau \right]^2 dt, \right. \qquad (22)$$

where $\delta > 0$ is considered to be unknown.

We shall seek the approximation of Z_t^{δ} to $\overline{Z}_t$ in the valuation of functional space $W_2^{'}[-1,1]$:

$$\|V\|_{W_2^{'}}^2 = \int_{-1}^{1} |V(t)|^2 dt + \int_{-1}^{1} \left| \frac{dV(t)}{dt} \right|^2 dt.$$

We shall consider below some specific test examples illustrating the workability, sensitivity, and features of the proposed algorithm.

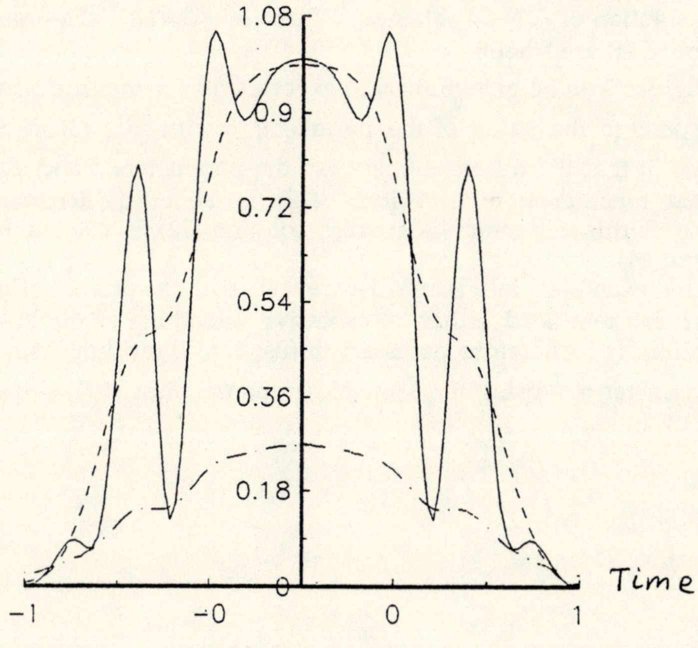

Figure 2. Influence of parameters δ and δ^2 on Z^{δ} solution by M=N=100, $\delta^2 = \delta = 10^{-3}$, ε=h=0.0 (See Figure 1 for symbols).

Numerical Test Calculations
Test Problem No. 1

We shall investigate the sensitivity of the solution of (22) to an inaccuracy in setting the error parameter δ>0. We performed the calculations for various values of δ and δ^2. Their results are presented in Figures 1-4. In these subsequent figures, the following notation is used:

W,N is the dimension of the matrix $\{K(t_i, \tau_j)\}$ corresponding to nucleus (19) of Equation (17); $\overline{Z}(t)$ is the exact solution of (18) (it is needed for illustrative comparison with the solution of (22)); $\tilde{U}(t) = \overline{U}(t) + \xi(t)$ is the right-hand side of (17) perturbed by an outlier and corresponding to the experimental results; (t) is a random function corresponding to (21), and is an interruption breakdown parameter. The random quality in (21) was determined as follows: θ is a random quantity from the sensor of random quantities on the segment [0,1] with a uniform distribution; δ_0^2 and δ_{int}^2 are the dispersions of the random processes $?_1$(t) and $?_2$(t) in (21), respectively.

Figures 1 and 2 give solutions of problem (22) with complete agreement of δ and δ_0^2 for ε=0 (no interruption in the observations). We can see that when the dispersion of the outlier δ_0^2 is greater at $\delta_{int}^2 = 0$, a smoother solution of (22) is obtained. When $\delta_0^2 = \delta = 10^{-3}$ (Figure 2), all the extrema on Z°(t) are smooth.

Figures 3 and 4 give numerical experiments for the case when δ_0^2 does not correspond to the value of the parameter δ with $\delta_{int}^2 = 0$. It can be seen (see Figure 3) that the difference between the parameters δ and δ_0^2 tolerated for normal functioning of algorithm (22) is about a decimal order of magnitude. If this difference is greater, solution (22) is excited by external noise (Figure 4).

The examples in Figures 1-4 reveal that a priori setting of the parameter δ>0 may lead either to excessive smoothing of the solution or to its oscillation. It is therefore necessary to use the algorithm with appraisal of the parameter δ>0 using the same experimental data $\tilde{U}(t)$, t∈ [-1,1].

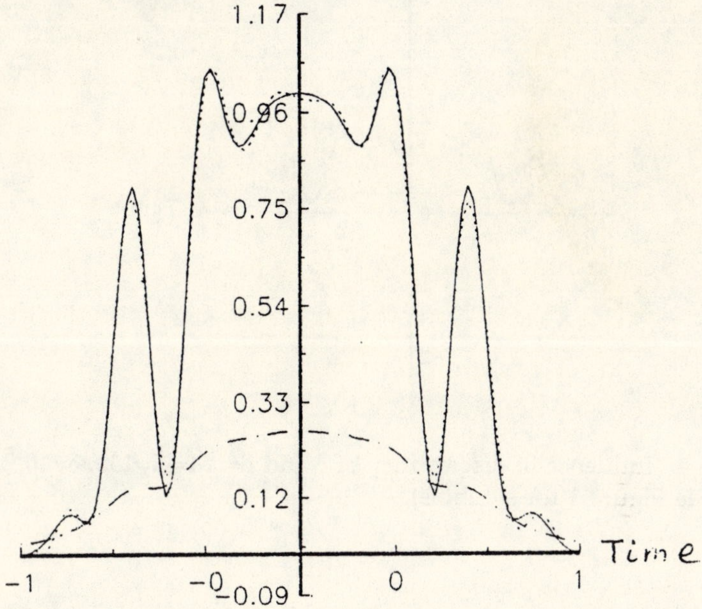

Figure 3. Influence of discrepancy of δ and δ^2 on Z^δ, $\delta=10^{-5}$, $\delta^2=10^{-4}$, ε=h=0.0 (See Figure 1 for symbols).

Test Problem No. 2

We study the influence of the appraisal of the maximum likelihood $\hat{\delta}$ of the parameter δ>0 on the nature of the solution of (22).

A glance at Figure 5 clearly reveals that when the appraisal of the maximum likelihood is introduced, the renewed solution of $Z^\delta(t)$ is approachably closer to the exact solution Z(t) in a uniform valuation than the solution of $Z^{\delta_0}(t)$ from (22). However, in this experiment (as in the preceding one), interruptions in the observations are absent.

According to (9), at a certain given value of $\delta_0>0$, a random process (mismatch signal) is calculated:

$$\Delta U(t) = \tilde{U}(t) - \int_{-1}^{1} K(t,\tau) Z^{\delta_0}(\tau) d\tau \qquad (23)$$

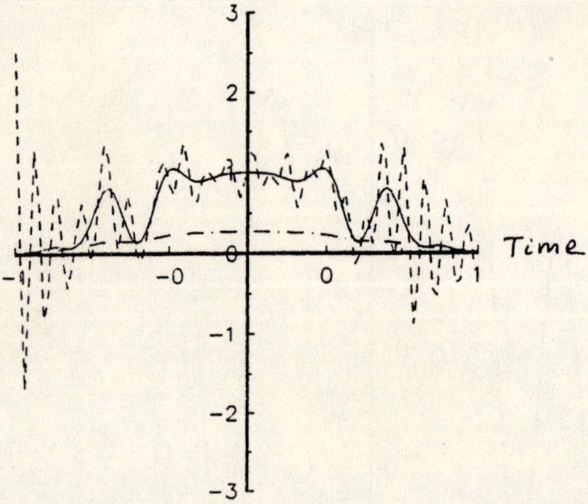

Figure 4. Influence of discrepancy of δ and δ^2. M=N=100, $\delta=10^{-5}$, $\delta^2=10^{-3}$, ε=h=00. (See Figure 1 for symbols).

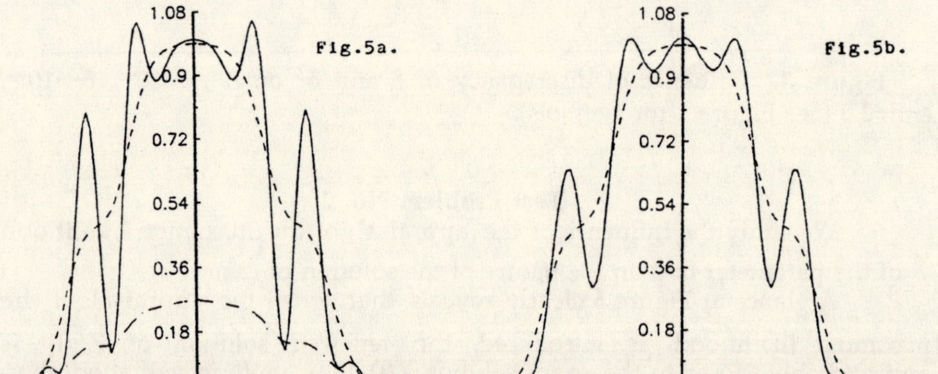

Figure 5. Maximum likelihood method - estimation of - parameter M=N=100,

$\delta_0^2 = 10^{-5}, \delta_{\text{int}}^2 = 0.0, \delta_0 = 10^{-3}$

$\hat{\delta} = 0{,}4698\, E-4\ \varepsilon = h = 0.0$

$a. — \overline{Z}(t), ---- Z^{\delta_0}(t), -.-.-.-.\tilde{u}(t)$

$b. — Z^{\hat{\delta}}(t), ---- Z^{\delta_0}(t)$

The realization of this process is shown in Figure 6 by the symbol ΔU_t. The figure also shows the functions $\tilde{U}(t) = \overline{U}(t) + \xi(t)$ and $U_t^{\delta_0} \equiv \int_{-1}^{1} K(t,\tau) Z^{\delta_0}(\tau) d\tau$.

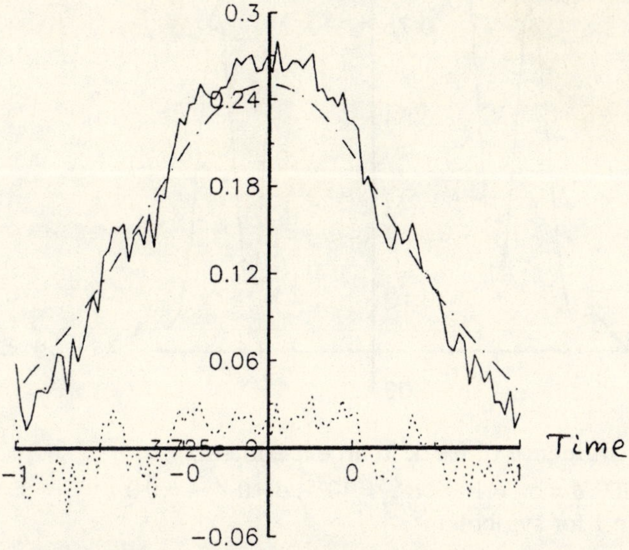

Figure 6. Maximum likelihood method - formation of mismath signal d0=10-2 (See Figure 5 for other parameters)
—$\tilde{u}(t)$, ---- $\Delta u(t)$, ---- $u^{\delta_0}(t)$

Test Problem No. 3

We study how interruptions in the experimental results affect the solution of (22) if we know the value of the error parameter $\delta > 0$ exactly, i.e. $\delta = \delta_0^2$ (hypothetically, this case has a low probability), but there is no information on there being interruptions in the measurements. In the given case $\delta = \delta_0^2$, and the outlier was simulated by (21) with various values of the parameter $\varepsilon \in [0,1]$. We adopted "sparing conditions" of interruption in the experiment producing no avalanche loss of stability. The random process determining the interruption was simulated from a Gaussian distribution with a zero mean and a dispersion close to the level of stability of the solution of (22).

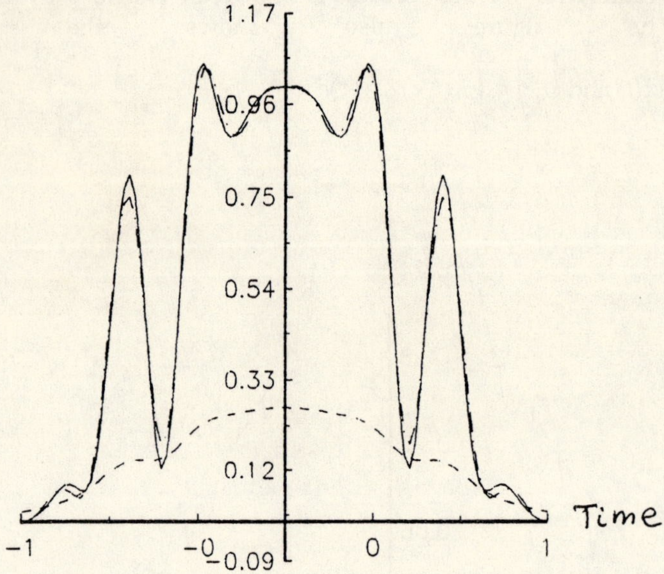

Figure 7. Influence of freak in an experiment
$M = N = 100, \delta = \sigma_0^2 = 10^{-5}, \sigma_{\text{int}}^2 = 10^{-3}, \varepsilon = 0,4, h = 0.0$
(see Figure 1 for symbols)

Figures 7-9 show the loss of stability of the solution of (22) because of interruptions when ε varied from 0.4 to 0.8, ε - is present number of the outliers in the data, δ_{int}^2 - is the dispersion of the outliers.

The data of Figures 7-10 allow us to conclude that even if the exact value of the error parameter is known a priori, the occurrence of interruptions in an experiment will nevertheless result in oscillation of the solution of (22). But it is impossible to determine the occurrence of interruptions in an experiment a priori. This is why algorithm (22) must be supplemented with an additional mechanism for automatically detecting interruptions in observations and their correction.

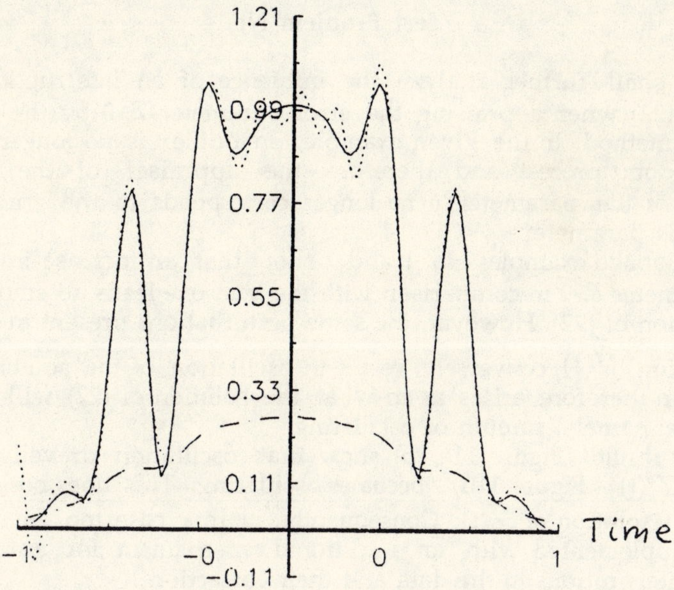

Figure 8. Influence of freak in an experiment
ε=0,6 (see Figure 7 for parameters and Figure 1 for symbols)

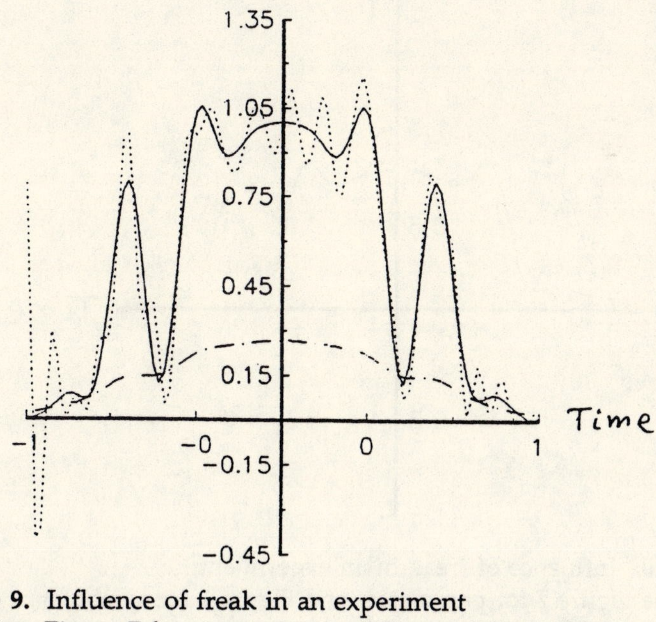

Figure 9. Influence of freak in an experiment
ε=0,7 (see Figure 7 for parameters and Figure 1 for symbols)

Test Problem No. 4

We shall further analyze the influence of an interruption on the solution of (22) when appraising the error parameter $\delta>0$ by the maximum likelihood method. In the given example, an outlier is no longer a steady normal random process and therefore the appraisal of the maximum likelihood of the parameter δ no longer corresponds to the true unknown value of this parameter.

The above examples No. 1 and 2 show that an increase in the value of the parameter $\delta>0$ in comparison with the true one leads to smoothing out of the solution of (22). However, the same perturbations present in the values of the function $\tilde{U}(t)$, conversely, result in oscillation of the solution of (22). The question therefore arises as to what the solution of (22) will be in the general case, namely, smooth or oscillating.

Our results (Figure 11a, b) show that oscillation prevails, and the solution of $Z°(t)$ - Figure 11b) - becomes oscillating. This does not correspond to the exact solution of $\overline{Z}(t)$. Consequently, in this case too the algorithm must be supplemented with an additional mechanism for automatically detecting interruptions in the data and their correction.

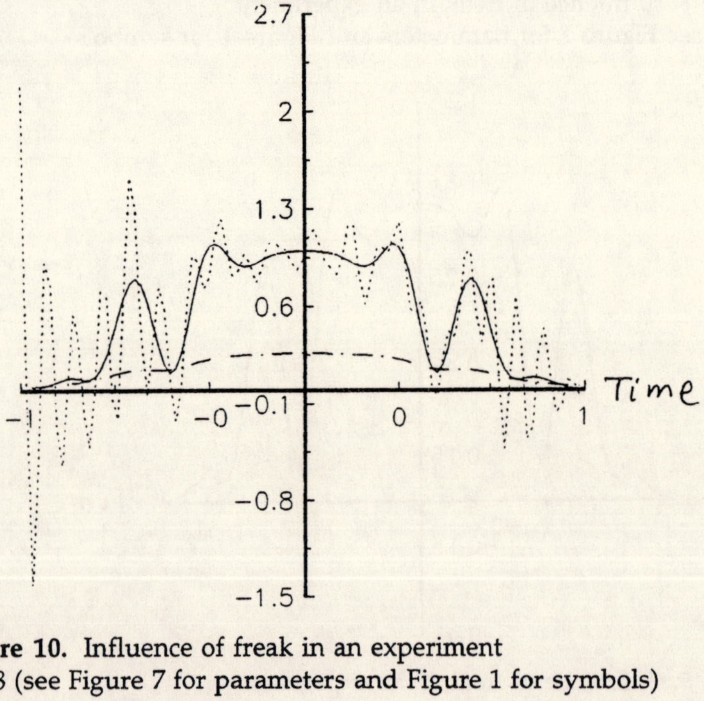

Figure 10. Influence of freak in an experiment
ε=0,8 (see Figure 7 for parameters and Figure 1 for symbols)

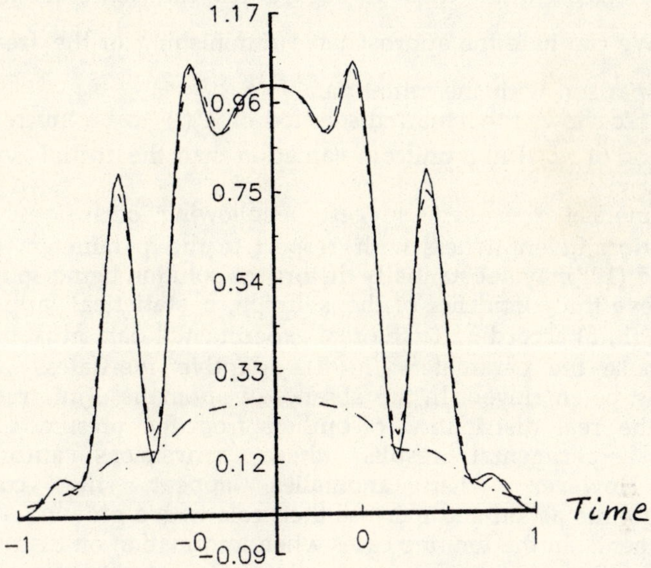

Figure 1. Influence of parameters and on solution by M=N=100, $\delta=\delta^2=10^{-5}$, ε=h=0.0. —$\overline{Z}(t)$,- - - $Z^{\delta}(t)$,—•—•—$\tilde{U}(t)$

Test Problem No. 5

Let us now consider the action of the mechanism for automatically seeking and eliminating inhomogeneities in the data $\tilde{U}(t)$. The algorithm of operation of this mechanism was described above in expressions (12)-(16). It consists in the automatic construction of an adaptive nonlinear inertialess converter that is optimal in the sense of (12). It makes the initial unsteady process steady, but never smooths it, and consequently, does not result in a loss of the high-frequency components of the initial data.

Figure 12a presents the initial data of the problem $\tilde{U}(t)$ in which, apart from the additive outlier, a freak value at the instant t≈-0.6 can be seen quite clearly. This freak value "oscillated the solution in Example 4.

In Figure 12b we can clearly see correction of the mismatch signal by the adaptive converter. Here ΔU_t is the mismatch signal without correction, ΔU_t is the same after correction. We can see that the adaptive converter automatically discovered a freak value at the instant t≈=-0.6 and reduced this value to one converting the entire random process into a steady one.

The corrected initial data in Figure 12a are represented by the function $\tilde{\tilde{U}}(t)$. We can note the approachable diminishing of the freak value on $\tilde{\tilde{U}}(t)$ in comparison with the initial data $\tilde{U}(t)$.

Figure 12c shows the renewed solution of $Z^{\circ}(t)$ to be much closer to the exact solution of $\overline{Z}(t)$ in a uniform valuation than the initial solution of $Z^{\circ}(t)$.

Our numerical experiments suggest the following conclusions:

1. The a priori indefiniteness with respect to the parameters (h, δ) in problem (8) and (12) may substantially distort the solution being sought.

2. To improve the confidence of the solution, a statistical appraisal of the parameters (h, δ) according to the real experimental data must be used.

3. To appraise the parameters (h, δ), adaptive inertialess non-linear conversions must be employed. In the absence of anomalies (interruptions or deviations in the real distribution of outliers from the presumed Gaussian one) in the experimental results, these conversions automatically degenerated. However, when anomalies appear, the conversions automatically go into action and increase their role with a growing number of anomalous elements. In the limiting cases when information on distribution of outliers is absent, the algorithm becomes a sign (nonparametric) one.

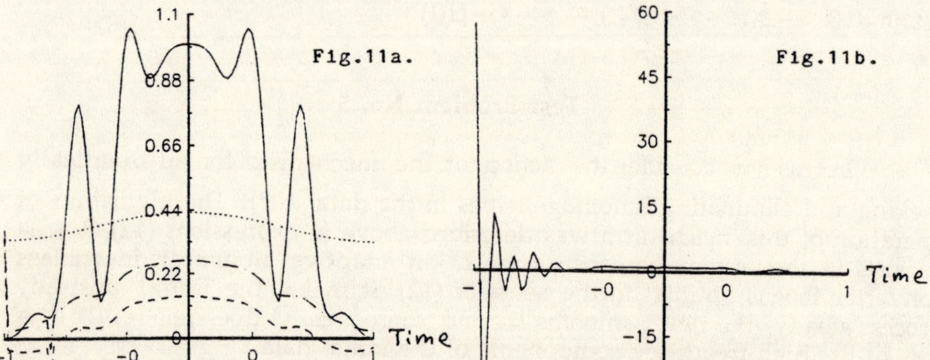

Figure 11. Influence of freak in an experiment on maximum likelihood estimation of δ-parameter

$M = N = 100, \delta_o = 0{,}1, \sigma_0^2 = 10^{-5}, \sigma_{\mathrm{int}}^2 \varepsilon = 0{,}01$

$\hat{\delta} = 0{,}5762\,E-02, h = 0.0$

$a. — \overline{Z}(t), ---- \Delta u(t), ---- \tilde{u}(t), \bullet\bullet\bullet\bullet\ Z^{\delta_o}(t)$

$b. — Z^{\hat{\delta}}(t)$

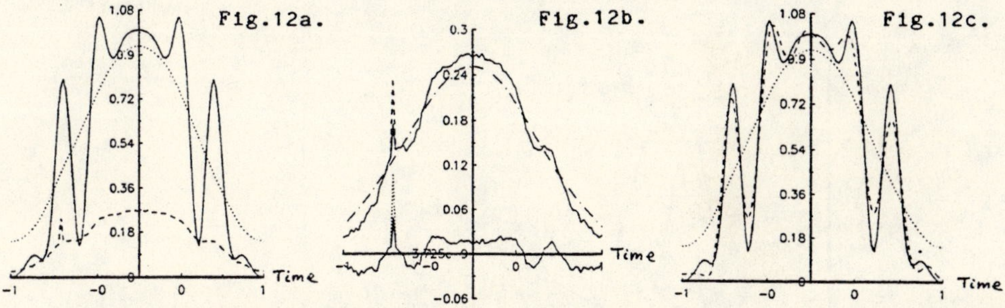

Figure 12. Correction of data by uninertia nonlinear converter

$\delta_o = 0,01, \hat{\delta} = 0,3986\,E-3$ (for parameters see Figure 11)

$a.$ — $\overline{Z}(t)$, - - - - $U(t)$,...... $Z^{\delta_o}(t)$;

$b.$ - - - $\tilde{U}(t)$, — $\tilde{U} = U^{\delta_o} + f(\Delta U) = f(t)$ (upper line),

___ $\Delta V = f(t)$ (lower line);

$c.$ ___ $\overline{Z}(t)$, - - - $Z^{\hat{\delta}}(t)$,... $Z^{\delta_o}(t)$.

Use of the Minimax Method in Solving Practical Problems

Let us now consider an example of using the algorithm (22) proposed above for calculations of specific material functions of a real viscoelastic medium. It was convenient to choose polybutadiene (PB), a linear flexible-chain polymer as the material simulating such a medium. The rheological properties of this polymer were described many times in the literature (see, for example, Ref. [2]). Moreover, in Ref. [14], the method of regularization was employed to calculate the function H(τ) for exactly this polymer.

We shall select the function of the relaxation modulus G(t) (See Equation 1) as the material function being determined, i.e. calculated with the aid of the above algorithm. The experimental results obtained in a rotary viscometer "Rheotron" of the firm Brabender were the initial ones for the calculations. The experiments were run at different gradients of the rates of deformation $\dot{\gamma}$ and the values of the shear stress τ corresponding to them were determined.

Hence, the object of the numerical calculations was to calculate the material function G(t) according to the preset function of the gradient of the deformation rate $\dot{\gamma}(t)$ and to the function of the shear stress determined experimentally.

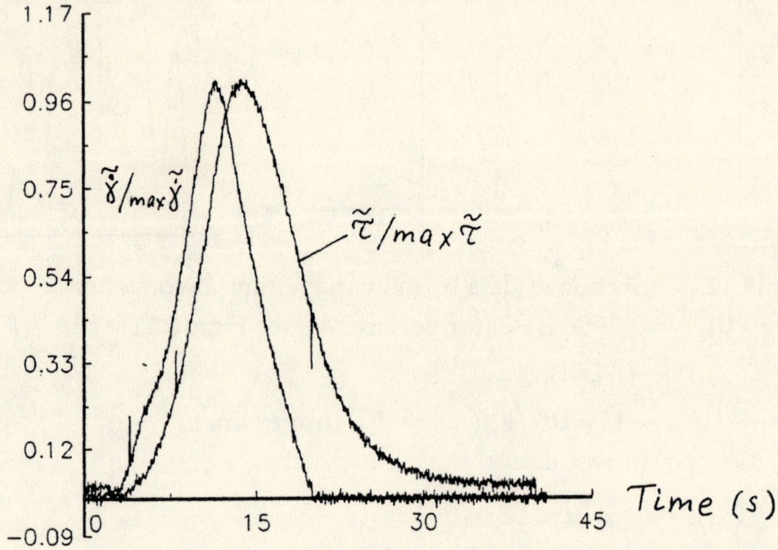

Figure 13. Experimental dependencies of $\dot{\gamma}(t)$ and $\tau(t)$ for polybutadiene (PB), M=8,3·10^4.

Figure 13 presents the normalized experimental functions $\dot{\gamma}(t) = \dot{\gamma}(t) + \xi(t)$ and $\bar{\tau}(t) = \tau(t) + \delta(t)$, where $\xi(t)$ and $\zeta(t)$ are random processes. [Normalization of $\tilde{\dot{x}}(t) / \max \tilde{\dot{\gamma}}(t)$ and $\tilde{\tau}(t) / \max \tau(t)$ was performed to visualize the outliers in the initial data] In the plots in Figure 13, we can see very clearly the anomalous freak values.

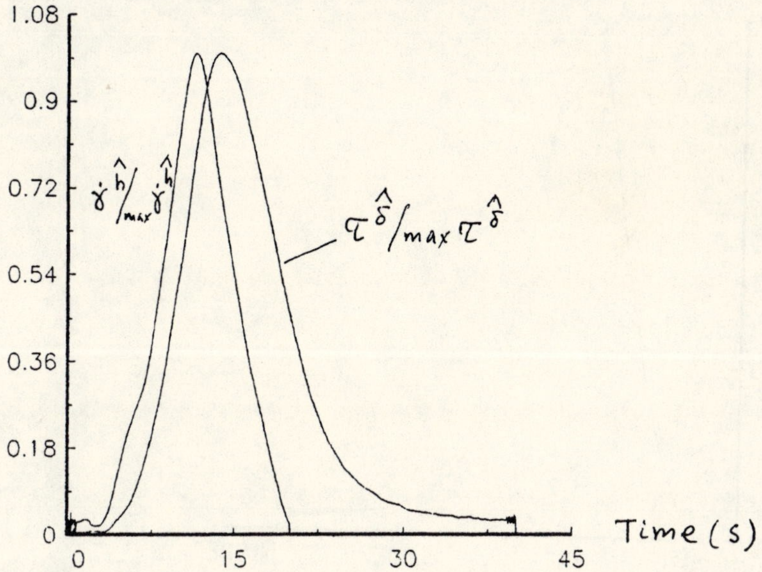

Figure 14. Dependencies of $\dot{\gamma}(t)$ and $\tau(t)$ for PB $M=8{,}3\cdot10^4$ after processing with the algorithm (22).

Figure 14 shows the same functions, but after processing with the described algorithm (22). Since the function $\dot{\gamma}(t)$ is set a priori, its mismatch signal was evaluated by the direct difference $\xi(t)=\tilde{\dot{\gamma}}(t)-\dot{\gamma}(t)$ and was processed by the above algorithm for seeking and eliminating anomalous freak values. Moreover, the operation of smoothing a random mismatch signal after eliminating the anomalous freak values was applied to the functions $\tilde{\dot{\gamma}}(t)$ and $\tilde{\tau}(t)$.

Figures 15 and 16 present the nonnormalized functions $\dot{\gamma}^{\hat{h}}(t), \tau^{\hat{\delta}}(t),$, respectively, used to find the function of the relaxation modulus $G^{\hat{\gamma}}(t), \hat{\gamma}=(\hat{h}, \hat{\delta})$ according to the above algorithm (Figure 17).

The function G(t) determines the initial data for calculating the relaxation spectrum of polybutadiene (See Equation 1) with the aid of a linear integral operator

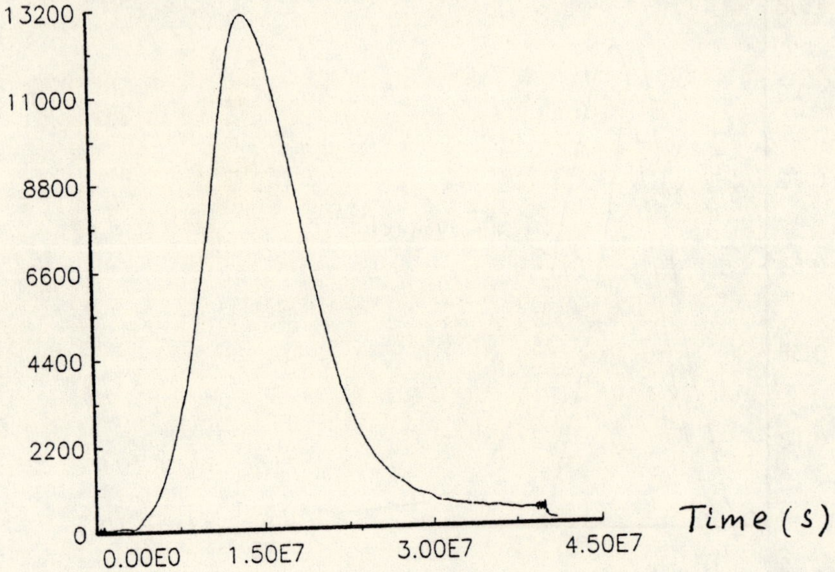

Figure 15. Dependence of stresses (Pa) vs. time (s) for PB, M=8,3·10^4, N=M=1024.

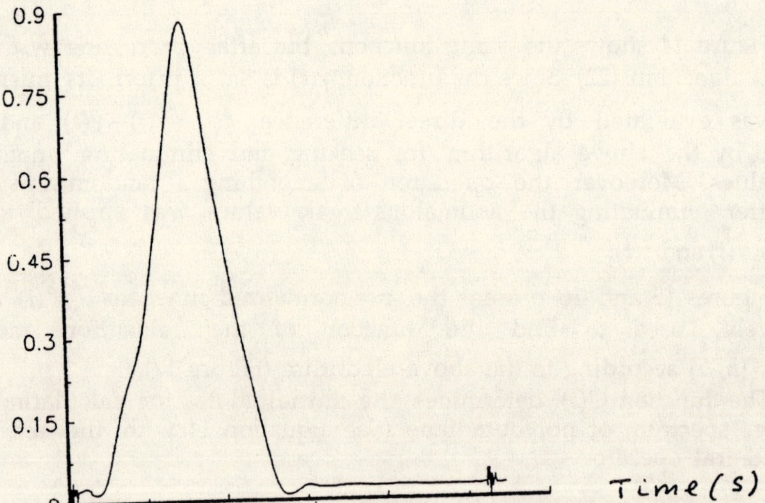

Figure 16. Dependence of deformation rate (s^{-1}) vs. time (s) for PB, =8,3·10^4.

$$G^{\hat{\gamma}}(t) = \int_0^t \exp\{-t/\tau\} H(\tau) d\tau \quad (23)$$

The function H(τ) is shown in Figure 18. It should be noted that the calculation program (Figures 15-17) employs the rapid Fourier transform and can process arrays of a considerable size. The program for calculation of the relaxation spectrum according to the relaxation modulus G(t), modulus of elasticity G'(ω) or loss modulus G"(ω) operates with an unsteady nucleus of the integral equation. This is why the algorithms of the rapid Fourier transform cannot be applied here, and the rapidity of program operation is based on the algorithm of rapid conversion of the initial matrix - nucleus to a tridiagonal form.

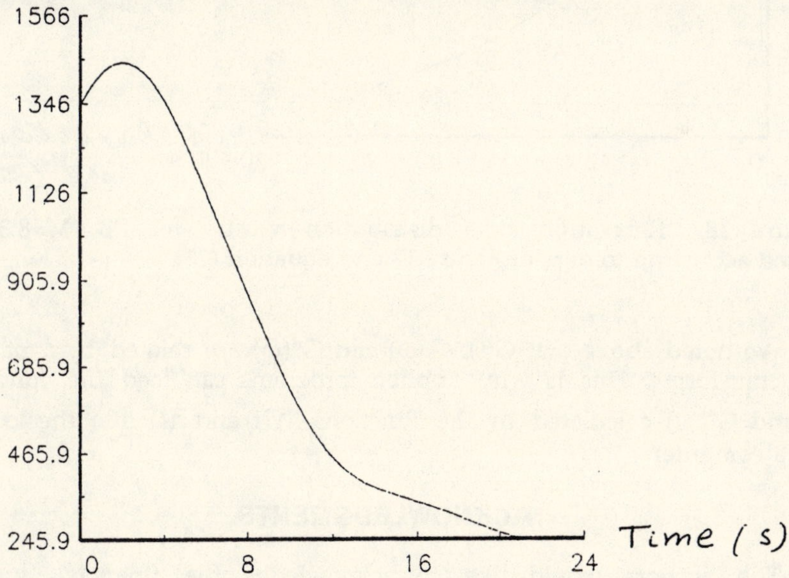

Figure 17. Dependence of relaxation modulus (Pa) vs. time for PB, $M=8{,}3\cdot10^4$.

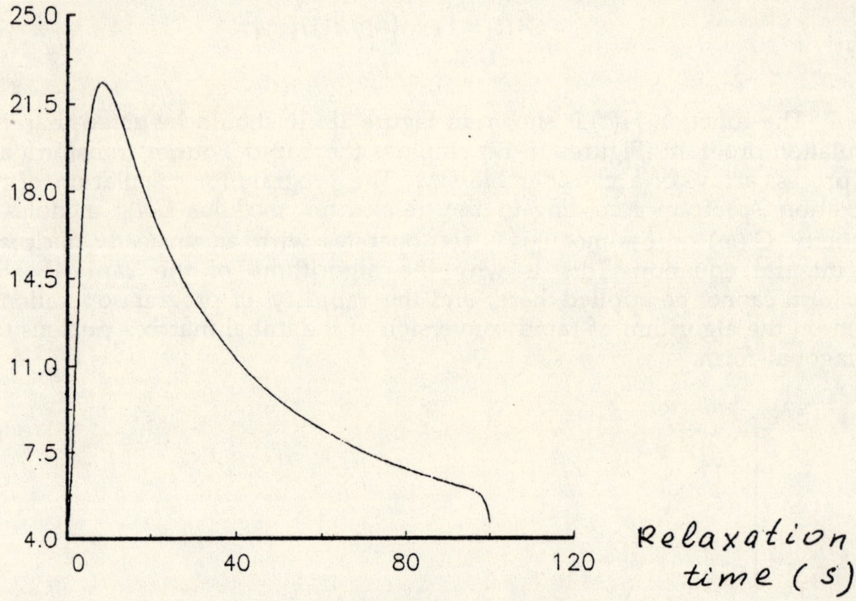

Figure 18. Relaxation time distribution spectra for PB, M=8,3·10^4, calculated according to date of Figure 17 and equation (23).

We noted above that G(t), G'(ω) and G"(ω) are related by a pair of Fourier transforms. This is why applied programs can feed the functions G'(ω) and G"(ω) calculated by the functions $\tilde{\gamma}(t)$ and $\tilde{\tau}(t)$ into the output files of a computer.

ACKNOWLEDGMENTS

The authors would like to acknowledge the financial support provided by Russian Found of Fundamental Investigations for support of this research through a research grant.

REFERENCES

1. J.D. Ferry, *Viscoelastic Properties of Polymers*, 3rd Ed., John Wiley Inc., N.Y., 1980.

2. Yu. Yanovsky, *Polymer Rheology: Theory and Practice*, Chapman and Hall, London, 1993.

3. D.D. Joseph, *Fluid Dynamics of Viscoelastic Liquids* Springer-Verlag, N.Y., 1990.

4. J.P. Royston, *Appl. Statist.*, 31, 115 (1982).

5. W.H. Press, B.P. Flannery et al., *Numerical Recipes*, Cambridge University Press, Cambridge, 1989.

6. C. Elster and J. Honerkamp, The role of the error model in the determination of the relaxation time spectrum, *J. Rheology* 36 (5), p. 911-927, 199.

7. J.W. Tukey, The future of data analysis. *Ann. Math. Stat.*, v. 33, pp. 1-67, 1967.

8. A.N. Tichonov, About Solution of Incorrect Problems and Regularization Methods, *DAN USSR*, 1963, 151, N3, p. 501-504.

9. V.A. Morozov, *Methods for Solving Incorrectly Posed Problems*. Springer Verlag Berlin, 1984.

10. C. Elster, J. Honerkamp, Inferring a retardation spectrum from creep data by a modified maximum entropy method. *Third European Rheology Conference*, Edinburgh, 1990, p. 147-149.

11. Yu.A. Basistov, A minimax solution of an equation of the first kind, *Soviet. Math. Dolk.*, vol. 27 (1983), N 1.

12. P.J. Huber, Robust estimation of a location parameter, *Ann. Math. Stat.*, 1964, 35, pp. 73-101.

13. Yu.A. Basistov, V.G. Tugushi, Regularized Algorithms of the Radiolocation Signals Detection, *Izvestia Vuzov, Radioelectronics*, 1987, 30, 4, 8-16.

14. C. Elster, J. Honerkamp and J. Weese, Using regularization methods for the determination of relaxation and retardation spectra of polymeric liquids, *Rheol. Acta*, 1991, 30, 261-174.

Elastomer Silicon Organic Heat-Shielding Materials For Aircraft External Surfaces

A.A. Donskoy
All-Russian Institute of Aircraft Materials
Moscow 107005, Radio st. 17

INTRODUCTION

Creation of engines for rockets and progressively larger sizes and aircrafts of ballistic and glide types require new heat-shielding materials (HSM) for exploitation under various conditions.

HSM of ablation type should operate at temperatures, a fortiori higher than the temperature of their thermal decomposition, and when the transition from one physical state to another occurs. Besides, HSM should be technologically efficient, light, resistant to atmosphere, vibrations and shocks, moisture and frost, either, have low heat and temperature conductivity, low coefficient of thermal expansion.

The rigidity of engine and aircraft casings being increased, HSM should be elastic not to be destructed at the deformation of protected constructions under mechanical and heat loads.

The problem of creation of elastic heat-shielding materials (HSM) was put forward in USA in early 60s [1] and in a relatively short time the covers on elastomer, including silicon-organic, basis appeared wide spread.

The work at the materials was carried out in the following main ways: creation of rubbers or rubber-like materials with mineral powder or fibrous fillers or their combinations, rubber-texture ones, reinforced by mono- or multilayered textures (like asbestos and glass-plastics), syntactic or chemically foamed materials, in a number of cases reinforced by combs. Besides, some materials worked out for other purposes, hermetics, for instance, were used as heat-shielding covers and materials.

MONOLITHIC MATERIALS OF GENERAL DUTY

Heat-shielding properties of silicon-organic rubbers were studied in detail in [2]. When high-temperature gas flows or plasma affect these rubbers on their surface there appears a carbonized precipitate. Depending on their composition the output can differ and the precipitate can be either strong or destroyed easily. Their use is of success at low and moderate heat flows. Silicon-organic elastomers being of "carbonized at thermal destruction" type, the materials on their basis are applied both as active and passive heat shield. The heat fed to the cover is adsorbed by volatile ingredients and low-molecular fractions is destructed elastomer. The formed carbonized precipitate serves as isolating medium, shields heat and radiation.

The principle of silicon-organic rubber operation is shown at Figure 1. Temperature distribution across the cover thickness shows, that resulting from the intensive heat adsorption in zones of melting, coke formation and thermal decomposition, the temperature difference may reach 1390° between the cover surface and the border line of decomposition zone and unchanged material. If methyl groups in polysyloxane are substituted by phenyl ones, the removal of the mass of the material is decreased significantly under the effect of super-sonic gas flow and heat flow, perpendicular to the specimen surface, 7900kJ/m^{2}s. The decrease was observed when the phenyl group content in the elastomer did not exceed 50%. After that erosion stability remains constant.

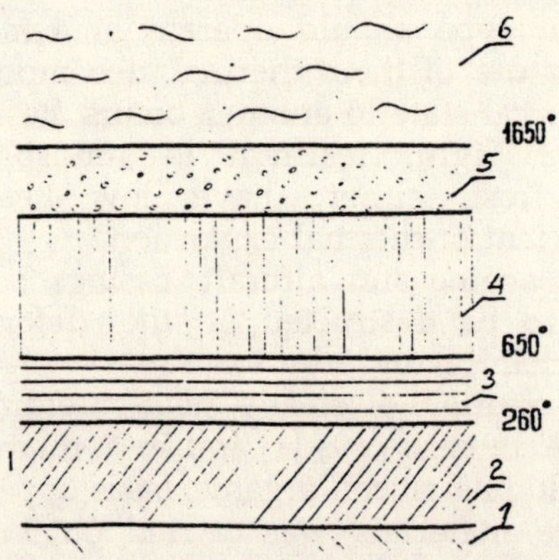

Figure 1. The principle of silicon-organic rubber operation.
1 - surface under protection,
2 - unchanged material,
3 - decomposition zone,
4 - porous coke,
5 - melting zone,
6 - gas flow.

A very interesting study was undertaken elsewhere [2] to evaluate the possibilities for substitution of carbon, introduced into polysyloxane elastomer as phenyl groups, by carbon black (high-disperse filler).

Studies in sub-sonic gas flow with enthalpy of friction equal to 8920 and 14450 kJ/m^{2}s and heat flow to the cold wall of 2260 kJ/m^{2}s have shown that carbon black filled materials appeared much more effective than those based on phenylmethylpolysyloxane elastomer.

Specialists try to find fillers which allow to get strong glass-like precipitate with high carbon content and increase the durability of materials before the beginning and after the finish of the ablation process. The latter aim is reached when heat-resistant powder and fibrous fillers are used. The decrease of density by foaming or introduction of light fillers appears rather interesting.

Materials for external heat-shielding of cosmic vessels (CV) must be resistant to vacuum and radiation influence. Materials based on silicon-organic elastomers are above the most of radiation-resistant organic materials as they contain no plastifiers and other volatile substances. Phenylmethylsyloxanes are more resistant to simultaneous influence of high-energy particles and high vacuum and less resistant to UV-irradiation than dimethylpolysyloxanes.

When high temperatures and high vacuum affect simultaneously, mainly low-molecular fractions evaporate, the rate of elastomer weight losses increases extremely sharply with temperature increase. So, the polymer specimen has lost 0.1 weight % during two weeks at 130°, while at 100° the same losses have been observed in a year. Silicon-organic based materials appear the best from elastomer ones by their resistance to radiation at normal temperatures in vacuum, their characteristic changes being extremely low.

Silicon-organic rubber HSMs are being studied in detail by many companies in USA and in less degree in Great Britain, France and Germany [3-9]. The effective ablation heat for silicon-organic rubbers was shown to be by 8 times higher, ablation rate by 7 and weight loss by 4 times lower than those for certain heat-shielding plastics. When the flame of oxygen acetylene burner having 3040° temperature was affecting a specimen 6.2 mm thick during 1 min, the temperature of the metal support increased just to 80°. Plasma stream with 4M rate, temperature 5000° and pressure 0.2 Torr (heat flow ~ 1220 kJ/m^{2}s), affecting rubber specimens, increased the surface temperature up to 1910°, a strong coke-like layer being formed. The ablation rate of the RTV-88 rubber (General Electric Company) reached just 15% from that of erosion-stable phenole-nylon composition. The SE-555 rubber decomposed even slower. The temperature was measured 8.5 mm from the surface of the RTV-60 rubber specimen, affected by plasma stream with the temperature 5000°. It was shown that in 6 min it increased just to 240°. Under these conditions a firm coke-like layer was formed at the rubber surface, the rest remaining unchanged.

Besides RTV-88 and SE-555 rubbers General Electric Company has worked out a lot of various silicon-organic materials, applied for heat shielding. To protect the second and the third steps of "Saturn" rocket from the flame stream with temperature 2760° (heat flow 1370 kJ/m^{2}s) for 10 min a material was used, cold-resistant down to -185°, vibration-sustained with 150 g acceleration and 3.2 mm amplitude and high-adhesible to the protected surface [10]. The material is ignitable and moisture-resistant, but has low rigidity. It has a high ratio of coefficients of emission and absorption in the wavelength region 0.5-2.5 mm. The material is technologically efficient and the repair of the damaged cover is not problematic. To protect steel strengthening rings for combustion chambers of this rocket engines RTV-602 rubber is used with relatively low density, and for the external detector RTV-11 rubber [11,12]. The detector cover during acceleration period is affected by blow waves and temperatures near 5000°. Heat-shielding of power rings of combustion chambers of six engines of both steps weighs 90 kg [13]. RTV-500 material is recommended for heat flows from 17 up to 13408 kW/m^2. Besides the material is efficient down to temperatures about -160°.

Besides "Saturn" large amount of silicon-organic rubber is used in other rockets, either. So, just for external surfaces of "Polaris" rocket engines 45.3 kg of rubber are used [14,15].

Tables 2 and 3 show some properties of non-vulcanized and vulcanized silicon-organic heat-shielding materials of General Electric Company [16-18].

Table 2. Properties of heat-shielding non-vulcanized materials of General Electric Company

Material	Color	Viscosity Pz	Consistence	Storage duration, months
RTV-77	white	8000	Paste	–
RTV-88	red	10000	Paste	–
RTV-90	red	12000	Stiff Paste	–
RTV-511	white	200	Liquor	6
RTV-560	red	400	Liquor	4
RTV-577	white	6000	Paste	4
RTV-580	red	8000	Paste	3

Table 3. Properties of vulcanized HSM of General Electric Company

Material	Density kg/m^3	Tensile strength kgs/cm^2	Stretch at break %	Rigidity by Shor (scale A)	Cold-resis-tivity ^{o}C	Contr-action %	Heat conduc-tivity, W/m·grade
RTV-77	1330	35.0	220	50	-67.8	0.3	-
RTV-88	1470	53.0	110	65	-67.8	0.3	-
RTV-90	1470	53.0	190	60	-67.8	0.2	-
RTV-511	1180	24.6	180	45	below -100	-	0.26
RTV-560	1420	56.0	160	60	below -100	-	0.31
RTV-577	1350	33.6	180	50	below -100	-	0.31
RTV-580	1490	56.0	110	65	below -100	-	0.31

Besides the above-mentioned materials the one named RTV-209 operates successfully in the flows with temperature 3315° [19]. Covers are made from it, designed for heat shielding of planes' trap doors and for heat-isolation of heat pipelines.

Sprayed cover was tested successfully at the X-15 plane with flight velocity 6800 km/h and the flow temperature 1090° [20].

English filial of General Electric Silicons has worked out a binary tixotropic germetic TBS758 for aerospace technology [21-23]. It is a white tixotropic paste well adjustable onto any support - metals, glass, plastics, wood, etc. Vulcanizate density is 1300 kg/m^3. The material is effective at temperatures 2700-2760°C for heat protecting and shielding during rockets' launching. This material is used also in space technology and for fire protection in ships, cars, buildings, etc.

To protect returning space vessels (SV) a rubber-like material [24] was worked out, which does not change during sterilization. It bears without cracking both heat blows and erosion effects during entering the atmosphere. In this respect it exceeds phenol plastics, graphite and ceramic covers. Non-electroconductive coke-like layer is formed on the surface of this material thus preventing from making windows for telemetric antennas situated inside the vessel.

Methods of General Electric for studying HSM are described in [25, 26]. The oxygen-acetylene burner (at 2705° during 60±1 s) imitates the conditions of CV entering the atmosphere, the oxygen-kerosene one - liquid reactive engine (LRE) jet systems operation. The characteristics shown in Table 4 are got at these devices. The results are greatly affected by the technology of the specimen formation.

Table 4. The results of fire-tests of the material of RTV-500 series

The material	The specimen reserve side temperature * °C	Ablation Rate	
		linear mm/s	weight g/sm^{2}s
RTV-511	28	0.0195	$2.34 \cdot 10^{-3}$
RTV-560	50	0.0160	$2.24 \cdot 10^{-3}$
RTV-577	26	0.0172	$2.34 \cdot 10^{-3}$
RTV-580	50	0.0122	$1.80 \cdot 10^{-3}$

*Specimen thickness not shown

Heat-shielding covers for wings and fuselage of XB-70 airplane worked out by General Electric Company, have high cold-resistance and adhesion to metals. About 1 ton of various rubbers on silicone-organic base are used in this place [27-29].

The company has patented a heat-shielding material based on polyorganosyloxane elastomer with silica filler and thin powder of a metal or metalloid (of silicium type) [30]. The material can be used as covers for rockets and space vessels returning to dense atmosphere layers. The general formula of polyorganosyloxanes has been worked out. It looks like

$R_a Sio_{\frac{4-a}{2}}$, where R can be a monovalent hydrocarbon radical or a part of

hydrogens can be substituted either by a halogen or by cyanalkyl radical. The elastomer viscosity can vary from 5 to 100 mln centistocks at 25°C. To vulcanize one can use benzoyl peroxide and radiation. Peroxide vulcanization needs heating. Germanium, tin and lead compounds (from metals) and those of silicium (from non-metals) are predominant as fillers. Arsenic, antimony and bismuth compounds can be used, either. The amount of a filler can vary from 0.01 up to 100 mass parts of elastomer, those from 2 up to 30 mass parts are predominant. The particles should be larger than 177μm.

Dow Corning Company has also worked out a large number of HSMs, based on silicon-organic elastomers. Its materials are used for "Polaris," "Minutemen," "Titan," "Saturn" rockets, "Jeminy" space vessel, "Crusador," "Bowing 727" and other planes [31]. These are materials, named Q-90-006, sylastics S-2048 and S-6510, 93-027, 92-009, etc.

Q-90-006 material is a binary system from the paste based on dimethylpolysyloxane and a vulcanizing agent, which allows to vulcanize without heating.

Sylastic S-6510 and S-2048 pastes are based on elastomers, containing small and large amounts of phenol groups, respectively, and their vulcanizing agents need heating for structurizing.

Erosion stability and heat-shielding properties of the materials can be evaluated by oxygen-acetylene burners, plasmotron Sg-102 of Plasmadin company (80 kW power) and Linde burner. Table 5 shows test conditions.

Table 5. HSM test conditions at set-ups of Dow Corning Company

Set-ups	Test Conditions		
	Fuel	Heat flow surface power, kJ/m²s	Gas flow rate
Oxygen-acetylene burner	Oxygen+acetylene	≤5080	low
Linde burner	Oxygen+kerosene	≤11320	1.2-3.0 M
Plasmotron	Nitrogen+oxygen	≤13540	subsound

Table 6 shows coefficients of efficiency of this company materials at heat flows and heat penetration rate, close to operating ones. The coefficient of efficiency is a dimensionless value, got by dividing of 100 by the product of the material density and heat penetration rate into the material mass. The higher this coefficient, the better operating properties of the material at the same test conditions are. The rate of that penetration is defined by the difference of the specimen thickness before the test and after removal of the coke-like layer. Test duration at heat flows 450 and 2960 kJ/m^{2}s was 60 s, and at heat flow 11320 kJ/m^{2}s - about 20 s.

Table 6. Heat-shielding properties of rubbers by Dow Corning Company

Material	Heat flow surface power, kJ/m²s	Heat penetration rate, mm/s	Coefficient of efficiency
Q-90-006	450	0.030	49
	2960	0.050	33
	11320	–	–
Sylastic S-6510	450	0.030	63
	2960	0.048	43
	11320	1.200	1.7
Sylastic S-2048	450	0.040	47
	2960	0.045	42
	11320	0.400	5.4

Table 6 shows that heat penetration rate increases with heat flow surface power increase. The latter being 11320 kJ/m^{2}s, the best results were obtained for S-2048 rubber based on elastomer with the highest content of phenyl groups. S-6510 rubber is less efficient and Q-90-006 material is inefficient at all. The higher is the content of phenyl groups, the higher is coke-like sediment durability and the cover works better at high gas flow rates.

Q-90-006 material besides good heat-shielding characteristics has a number of other important properties. It allows to use it for plane and rocket deflectors [342].

The company has patented [33] a heat-shielding material for returning SV, containing mineral fillers and based on phenylmethylpolysyloxanes with (0.25÷1.25):1 ratio of phenylmethyl and dimethyl groups at silicon atoms.

Physical-mechanical properties of the above-described materials are shown in Table 7.

Table 7. Physical-mechanical properties of HSM by Dow Corning Company

Material	Density, kg/m^3	Tensile strength, kgs/cm^2	Lengthening at break, %	Solidity by Short (scale A)
Q-90-006	1480	38.6	150	50
Silastic S-6510	1210	70.0	400	50
Silastic S-2048	1320	39.5	300	60

Test results for some materials at plasma set-up at various values of heat flows and plasma enthalpies are shown in Table 8 [34]. These data allow to evaluate coke-like sediment effect on ablation intensity. Seeming heat of ablation was determined by the specimen weight loss and the value of heat flow directed to cold wall, irradiation and heat blocking effects and coke-like sediment surface temperature considered.

Table 8. Test results of elastomer materials in plasma flow.

Material	Plasma enthalpy, kJ/kg	Specimen surface temperature, °C	Heat flow, kJ/m^2 s	Seeming heat of ablation, kJ/kg solid sediment		Solid sediment properties
				with	without	
Vulcanized without heating:						
binary of low density	7670	1675	1126	143800	13380	Durable glass-like. High adhesion to the material
	13700	1790	3387	18650	12800	the same
binary of low density (cold stable)	7670	1650	1128	75750	13300	the same
	13100	1810	3387	18900	13380	Durable granules. Heat adhesion to the material.
binary with FeO_3	7670	1675	1128	96100	11580	Durable glass-like, low adhesion to the material.
	13700	1675	3387	7800	6820	the same
binary based on phenylmethylpolysyloxane	8650	1485	1187	21900	7780	Durable granules. High adhesion to the material.
	13100	1610	3387	18450	11500	the same
teflon (control specimen)	8650	1210	1187	–	3370	Not cokes. High adhesion to the material.
Vulcanized at heating:						
dimethylpolysyloxane elastomer	7670	1485	1128	63950	13800	Durable glass-like High adhesion to the material.
	13100	1650	3387	–	8950	Soot-like. No adhesion to the material.
phenylmethylpolysyloxane elastomer	10800	1595	1242	67500	11200	Durable glass-like. High adhesion to the material.
	13700	1620	3387	20250	36400	the same
Phenylmethylpolysyloxane elastomer with special fillers:						
modification 1	8100	1486	1187	75750	7780	Durable glass-like. Good adhesion to the material.
	13100	1675	3387	23200	15220	Very durable glass-like. High adhesion to the material.
modification 2	8100	1540	1187	46500	9450	Durable glass-like. High adhesion to the material.
	13100	1840	3387	16920	12870	Durable granules. Good adhesion to the material.

Under natural conditions enthalpy of the material decomposition can be decreased to a great extent because of vibration, blows, etc.

The materials of Dow Corning Company were also tested as heat-shielding for sounding rockets[34]. The covers of three modifications were tested:

1) made by molding and gluing to protecting surfaces without heating;
2) sprayed, vulcanized at room temperature;
3) made from preparations, shrunk at moderate heating (just for cylindrical surfaces).

Molding of great parts of covers and their attachment are rather complicated. That's why for preliminary tests sprayed materials were chosen. Their properties are shown in Table 9.

Table 9. Sprayed materials properties.

Material	Density, kg/m³	Tensile strength, kgs/cm²	Lengthening at break, %	Specific heat, kJ/kg. grade
92-009	1090	42	600	1.465
92-007	1540	21	250	1.465
92-027	870	35	50	1.340

The best cover surfaces were got by evaporation of 93-027 material cold vulcanized. However, 92-009 material was higher durability and elasticity, it being very important at land-based rocket operation. Fire tests of specimens from this material 0.30 and 0.76 mm thick on aluminum plates 3.2 mm thick were made at low (565 kJ/m^{2}s) and moderate (2260 kJ/m^{2}s) heat flows and moderate gas flow rate. Oxidizing gas flow was directed perpendicular to the specimen surface.

Tests have shown that 92-009 material has good ablation properties and is relatively insensitive to oxygen content in the gas flow. Difference in calculated and experimental data for low values of heat flows was negligible. At moderate heat flows this difference was significant, as changes of the cover efficiency in dependence on heat power were not considered during calculations. It was established that the material works adequately as a cover for head corner, sections, engine bodies and steering planes of rockets. It was used for heat-shielding of retro-rocket engine under rather difficult conditions of aerodynamic heating at rate power 12.2 kgs/cm^2 when entering the Earth atmosphere.

92-027 material is used to protect "Titan" rocket and X-15X-2 plane [35]. The cover is sprayed onto the undercoat in two steps. First, a 1.3 mm thick layer is applied which is being vulcanized during 16 hours at 21-45°. It provides good adhesion with the second layer, which is being applied consequently in dash-like manner until a preset thickness is reached (from 0.4 mm at the back edge of the wing to 7.6 mm at the front one). To provide mechanical damages' protection the external surface of the cover is coated by a layer of 90-060 silicon-organic resin of Dow Corning Company [36]. The external surface of solid fuel rocket engine (SFRE) of the first section of "Titan 3-S" launch-vehicle is shielded by a new elastic cover 4.6-6.3 mm thick. The material is applied to the metal as cast [37]. During a rocket flight at 1370° the external layer of the cover cokes.

To protect "Crusador" plane fuselage from gas flow of "Sidewinder" rocket, launched in flight, the material 2-0103 is applied successfully [38]. The material Q-3-0121 has the same application [39]. It contains cyclohexane and n-hexane. The cover is sprayed onto zincum-chromate undercoated with silicon-organic adhesive underlayer and vulcanized without heating. The erosion of the cover 1.25 mm thick was about 0.25-0.38 mm for every of three rocket launching at 1500 m height. It is increased if liquid fuel or fluid from the plane hydraulic system falls onto it. However, fuel- and oil-resistance of the material can be increased by substitution of dimethylpolysyloxane by fluorsyloxane elastomer.

The company has worked out heat-shielding rubber which has special high electroinsulating properties [40]. It is vulcanized at 150°, and is cold-resistant down to -73°C.

Devices within the first section of "Saturn" rocket accelerator are heat protected by a binary silicon-organic rubber vulcanized at room temperature [41, 42]. The cover from this rubber provides the temperature in the device compartment below 65°. The same material is used as an external rocket cover 0.76÷3 mm thick.

Heat-shielding properties at high-power heat flows can be improved sufficiently if silicon-organic elastomers with high content of vinyl links [43] are used as a base. The material content (in mol %): 40-45% of vinyl-containing elastomer with molecular weight 22000-24000, viscosity of 650-1690 centipuaz at 25-28°; 50% - of silicon filler, silicon carbide and carbon fiber; 5-9% - of polymethylhydrodimethylsyloxane and 0.25-0.75 % of platinum hydrochloric acid, containing 3 - 3.5% of platinum in cyclic vinylsyloxane. Thermal stability of vinyl-containing polysyloxane (427°C at 100% methylvinyl links) is higher than that of dimethylpolysyloxane (232°C) and increases along with the increase of vinyl group content in the elastomer. 100% methylvinylpolysyloxane has the minimal mass losses at 800°C which equal 18%. Dimethylpolysyloxane looses 10% of mass at 440°C, while 100% methylvinylpolysyloxane - at 520°C.

Fire-heat-shielding materials are being worked out, either. So, to protect "Bowing 727" from 15-minute and fuel tank of small reactive planes from 16-minute influence of flame with temperature 1095° a light rubber RTV-325 is applied [44, 45].

Michigan Nasil Corporation has also joined the process of elaboration of elastomer silicon-organic HSM, having worked out a material R-2757 [46]. Under the influence of open flame with temperature about 2760°C a solid cake is formed on its surface, which possesses heat-shielding properties for 60 s.

A heat-shielding cover based on silicon-organic elastomer [47] was elaborated by Grumman Aerospace Corporation to protect external surfaces of F-14 plane from exhaust of fuel combustion products of Sparrow rocket engine.

A spray-coating material was patented in BRD, which contains a mixture of boron and aluminum silicate ($3Al_2O_3.2SiO_2$) [48], as a filler.

The optimal formula consists of 2.5 mass parts of boron and 12.5 mass parts of aluminum silicate. This material contains also 37.5 mass parts of white-spirit.

The material got by the method patented by NASA (USA) [49] is also a spray-coating one.

The material includes (in mass parts): 75% of dimethylpolysyloxane, 10% of phenolformaldehyde microspheres, 11% of silica, microquartz filaments 6.2 mm long and 5 μm in diameter (the quantity not shown), 5% of polydimethylsyloxane oil and 50% of chlorinemethylene solvent. Vulcanizing system consists of 28% of alkaline catalyst and 72% of chlorinized solvent, for instance, chlorine-methylene. The technological process contains several steps. At first, 1034 g of microspheres and 940 g of silica are being mixed in vacuum (736 mm Hg) for 3 hours at 93°. Then the mixture is cooled and stored until application. To get the material a precise quantity of the mixture should be taken.

Elastomer and oil are mixed separately. Filaments are added to the mixture. Then it is remixed till homogeneity is reached. To get the material a precise quantity of this mixture should be taken. Then fillers and elastomer composition are mixed till homogeneous mass is got. The solvent is added, either, to get a pre-set viscosity. With the help of a sprayer it is then mixed with the catalyst solution.

Spray-coating silicon-organic covers are shown [50] to be applied successfully both for heat-shielding of space vessels and to protect elements of launcher. Moreover, such covers can be reconstructed many times both when used in launchers and to fix technological joints in SV.

Some silicon-organic rubbers can be used as heat-shielding and electro-insulating covers [51]. Such rubber has the density of 1.18 g/cm^3, coefficient of heat conductivity 0.07 W/m.grade and specific heat capacity 1.39 kJ/kg.grade. The rubber works well as a heat-shielding material at moderate heat flows.

To decrease the rate of silicon-organic composition carrying away it appears useful to reinforce them with carbon fabric or fibers, the content of reinforcement being below 20% [52]. Socket model tests made from the material in the flow of electric arc device (enthalpy of retardation 4620-8820 kJ/kg, pressure 1.5-3.5 kg/cm^2) have shown that reinforcing decreases the rate of carrying away more than twice, the heat conductivity and density being practically constant. The material was worked out for heat-shielding of a vessel, entering Venus atmosphere.

A heat-shielding material for protecting aircraft surfaces from intensive heating during 2 minutes [53] is a monolithic one. It is made as sheets 6.3 mm thick and consists from a double-layer fabric impregnated by liquid silicon-organic composition, a layer of silicon-organic rubber and an external reflecting aluminum layer. Refrazyl (SiO_2≡98-99%), spacequartz (SiO_2≡99.5%) and silica (91-93%) [54-56] can be used as fabrics, either.

Inflatable space vessels, returning to the Earth atmosphere, and space suits, providing astronaut protection when he leaves a damaged space vessel, appear an original area for elastic HSM application.

For inflatable SV metallic fabrics with covers based on silicon-organic elastomers were worked out [57]. So, the metallic fabric with elastic cover, containing crushed glass frit, can be used at temperatures up to 815°. Elastomer in the cover begins to decompose at 426°, the glass forming a viscous melt, which keeps solid products of decomposition and retains adhesion to the fabric. Glide SV are planned to be made from a metallic fabric (alloy containing 73% of nickel and 20% of chrome), impregnated by silicon-organic rubber [58]. During returning into the Earth atmosphere the main part of rubber is consumed because of ablation, the fabric temperature remaining below 540°.

MATERIALS OF LOW DENSITY

Much attention abroad is paid to materials of rather low density (<1.10 kg/cm^3). Two directions exist: 1) working out of foams with the help of chemical pore makers; and 2) using of light fillers, mainly phenolformaldehyde, glass, silica and other microspheres and tubular filaments. The materials with microspheres are called "syntactic foams."

The above-mentioned companies are creating foams. General Electric has worked out a large number of such materials. RTV-7 is a black fluid liquid composition with viscosity 7000-9000 cst. It is vulcanized without heating and makes an elastic foam [59]. The foam density can be varied by mixing RTV-7 with non-foaming fluid liquid compositions in various ratios. Vulcanization and foaming can be performed in various media, for instance, in water. RTV-7 can operate for a long time at temperatures from -54° up to 316° and is used mainly for heat protection. It can also be used as light electric insulators, shock and vibration absorbers.

RTV-757 is a tixotropic paste based on methylphenylpolysyloxane elastomer. It is spread onto a protected surface by hot air spraying. Simultaneously it is foamed, its volume increasing by 35% [60,61]. The material is being vulcanized for 5 min at 150° without pressure. The product has the density of 700-800 kg/m^3 and rough surface. Vulcanization occurs in the contact with various materials, except rubbers, based on butyl-caoutchouc, chloroprene, sulphur-containing compounds and amines. So, the underlayer from SS-4120 is recommended. RTV-757 can also be spread by pouring. Any method of spreading provides its high adhesion to the substrate at low and high temperatures. it is also cold- and heat-stable, has good ozone- and chemo-stability, dielectric properties, is low contracted at vulcanization and stable to aging [62]. Its tensile strength is above 17 kgs/cm^2, elasticity being kept at temperature -101°.

Powdered quartz or diatomite increase the density, while silicium dioxide, got in gaseous phase - the durability of the material (SiO_2 decreases heat stability to a certain extent, that can be increased with the help of special stabilizers).

The flame of oxygen-acetylene burner with temperature 2760° forms a stable coke-like layer on the surface of RTV-757 specimens which protects them from future destruction. The temperature of cardboard protected by a thin layer of the material increased just by 8° during 60 s [59, 63-66]. The strips from the material with sizes 153x19x12.7 mm, hung in vertical heat-resistant glass tubes of 62.5 mm diameter, being affected by gas burner flame with temperature 1100°, remained elastic, did not crack and ignited during 20 s [63, 67]. Specimens 6 mm thick on a metal plate resisted the flame during 15 min. Thanks to high heat-protecting properties, RTV-757 material is used for heat protection of surfaces of planes, tanks, ships and rocket launchers [68].

Foams from the series ESM-1000 of the same company are designated for external heat protection of returning SV of gliding type [69-71], SV for flights to Mars and Venus [72-74], artificial Earth satellites [75], X-15-2 plane [76] and the system of astronaut resuce from SV [77].

SV, returning to the Earth, have non-symmetric configuration and vane surfaces for complex maneuvering. This makes it difficult to calculate aerodynamic heating and to choose heat-shielding. HSM moderate air flow perturbation effects, thus providing considerable freedom in choice of flight orbit and the vessel maneuverability. The design of heat-shielding of gliding SV is complicated by their long flight in the Earth atmosphere (1000-3000 s) at moderate heat flows and the necessity of pre-landing maneuvering with low rates, thus increasing the flight time by 500-1000 s more. At the orbit the temperature of the shade side of the vessel can reach -179°, and that on the illuminated side +95.

Heat-shielding covers for such SV should have the following extra properties:

1. Durable contact with the protected surface and physical-chemical characteristics should be preserved at temperatures from -180° up to 370° under long effect of vacuum, UV-irradiation and other factors in the space. They should have low density, modulus of elasticity and coefficients of heat conductivity and thermal enlargement.

2. The possibility of HSM spreading onto the surface of a complex configuration with minimal quantity of junctions and clamps.

3. The possibility to prepare covers with changing density and cross section to achieve their maximal efficiency and to provide necessary resistance to erosion effect of air flow at different places of SV surface.

4. High form stability and reliability at hand control of SV.

Phenol-nylon material was planned to be used for heat-shielding of gliding SV, however, the temperature interval of its operation is not wide enough. Testing of ESM-1000 series materials have shown that they fulfill the most requirements to HSM for these vessels. Materials of this series are coking compositions based on phenylmethylpolysyloxane elastomer with pore creator and after vulcanization have density from 320 to 1280 kg/m3.

The modulus of elasticity for foams does not exceed 14 kgs/cm^2 at temperature below -100°, heat conductivity at the density 368 kg/m^3 equals 0.038 W/m.grade (twice lower than that for phenol glass plastics, filled with refrasile), and specific heat capacity equals 1.67 kJ/kg.grad. Coefficients of linear expansion of ESM-1004 material and the metal are so close at temperature below -118° (brittleness temperature), that it appears very favorable for operating properties of the construction. Tests of metal and glass-plastic panels with ESM-1004 material at cyclic biaxial tension have shown that they retain efficiency down to -185° and those with ESM-1001 - down to -170°.

Materials from ESM-1000 series being tested in plasma flow during 1200 s (retardation enthalpy 11700 kJ/kg and heat flow 113 kJ/m^{2}s, directed to a cold wall), the linear carrying away for the materials with densities 370, 620 and 975 kg/m^3 was 2.6, 1.0 and 0.34 mm, respectively. For gliding SV with lift to drag ratio, equal to unity, the weight of 1 m^2 of heat-shield with the density 300 kg/m^3 is 34 kg at the head part and 24 kg at the back one. At the same time the temperature of the protected surface during gliding time in atmosphere does not exceed 100°. If the temperature can be increased up to 315° the weight of heat-shield can be decreased by 2-2.5 times.

ESM-100 materials can be modified to improve thermomechanical and ablation properties. Two ways exist: 1) increase of phenyl group content in elastomer to decrease its temperature of brittleness down to -118° without spoiling their heat resistivity; 2) introduction of special admixtures and fillers, i.e. asbestos, to increase durability and coke output.

To prepare HSM one should carefully mix fluid liquid elastomer with fillers, vulcanizing and foaming agents, pour the mixture into an open mold, vulcanize and foam at room or increased (50-60°) temperature for several hours. Taken out from the mold, the sheet is re-vulcanized at 150° for 10 hours. The ready material can be cut into sheets of given sizes or it can be formed as necessary by polishing.

Elasticity of these materials makes it easy to glue them to surfaces of complex foams. To do it one should carefully clean them, degrease and cover by adhesive underlayer based on phenylmethylpolysyloxane elastomer or a glue. The glue is vulcanized without pressure for 32 hours at room temperature and for 24 hours at 50°.

Foaming materials can be spread over the protected surface by various methods, depending on the form and accessibility of its parts. So, the revolution surface can be covered by a thick layer when sections are glued, the slots hermetized and machined to provide necessary size and smoothness. To protect bolding, at hatches, for instance, stoppers from the same material are used. They are glued into the holes in heat-shielding layer with the help of quick vulcanizing glue. The covers can also be formed by spraying with a brush or a roller. Spraying was used to deposit HSM onto X-15 plane [78]. For X-15-2 plane two materials from ESM series, namely ESM-1004-DR and ESM-1004-VR, were chosen [76]. In both cases the covers are formed as sheets, glued to the surface.

Two materials were made for heat-shielding of SV, intended for flight to Mars, orbiting it and landing. They are chemically foamed ESM-1004-X based on phenylmethylpolysyloxane, filled with inorganic fiber and ferric oxide, belonging to ESM-1004 group of materials with the density 0.24±0.003 g/cm^3, and chemically foamed ESM-1030 on the basis of epoxy-polysyloxane, also filled by inorganic fibers, density 0.29±0.05 [73]. Experiments showed them to keep their work-ability after sterilization by ethylene oxide or heating, and under fungi influence. The materials were tested under the conditions of the flight to Mars at cyclic changes of temperature from -185° up to 150° at appropriate tensions and deformations and under the conditions of landing to Mars, either. It appeared that at cyclic actions the material ESM-1004-X stands up to low temperatures down to -185° while ESM-1030 only down to -130°. Tests in high temperature flows proved them both efficient as heat-shielding systems.

The same materials were tested as heat protectors for the vessels, entering Venus atmosphere [72]. High temperature tests were made in gas mixture of 70% N_2 and 30% CO_2. The value of integral emission coefficient appeared to depend strongly on the material composition and the conditions of hard sediment formation. At the same time metal silicate introduction increases the emission ability in IR-range.

The materials ESM-1004-X and ESM-1030-1B (alumosilicates used as inorganic fiber) can be used unusually to rescue three astronauts from the space vessel [77].

Materials from this series are used for heat protection of artificial Earth satellites, either [75]. Besides good heat-shielding properties these covers must resist orbital cyclic temperature effect, which can vary from -115° up to 260°.

During returning of the satellite in dense atmospheric layers the head surface can have temperatures above 1927°, side ones about 1600° and the stern one 926°. One should note that the satellite enters the atmosphere by its stern and then at the height of about 84 km aerodynamic forces re-orient it, the flight being continued with the head part in front.

Close coefficients of linear expansion of the cover and the protected surface in a large range of temperature changes at the orbit is one more demand of heat-shielding of satellites. The orbit parameters affect the cover heating by solar irradiation, either. Maximal values of heat flow can be reached during the flight at low heights and large angles between move direction and solar irradiation, while at large heights and small angles not the whole surface is heated. Besides, the cover should be high-resistant to space factors (vacuum, UV-irradiation, temperature, action of micrometers and other particles, etc.), should not damage during ground servicing, should be easily repaired after damages and have sufficient operating time.

Foam ESM-1001PS (based on methylphenylpolysyloxane with asbestor fiber) appeared good for an aluminum construction under cyclic heat loads in temperature interval from -170° up to 149°, for a phenol glass -

plastic - from -157° up to 121°C and as an external cover of a stern part of experimental and exploited strategy SV, entering the atmosphere.

Asbestos fiber being substituted by shorter inorganic ones (from 6 to 20%) in the materials of this series, they acquired improved heat-shielding properties and erosion stability. The fact can be explained by formation of more durable cokelike layer, more stable to aerodynamic shear force effect. Besides, because of this substitution the durability of the materials and their compatibility with metallic constructions also increased. The materials turned more simple in production and exploitation. One of them - ESM-1004-PS at the aluminum construction was successfully tested in the cyclic temperature region from -184° up to 149°C. Varying fiber concentration in the elastomer, one can elaborate the materials, compatible with those of the protected construction at temperatures below the critical one of glassifying - 118°C. These materials were tested successfully during flight as covers for front edge, underfuselage stabilizer, aerodynamic brake, liquid oxygen tank and other surfaces of X-15 airplane.

Silicon-organic materials have a disadvantage in space application. They include low-molecular silicon-organic compounds which isolate from them in space and are condensed at colder or more polar surfaces. So, when necessary, specially prepared elastomers are used with minimal content of compounds volatile and satisfying special demands concerning space. Such elastomers can also vulcanize without heating, however, their cost is extremely high - about 882% per kg.

According to the tests in low-temperature plasma, the materials of ESM series are divided into 5 types:

1) materials, working in high heat flows at temperatures above 1617°C; here the surface carrying away occurs;

2) materials for heat flows with temperatures between 617° and 1617°C; no surface carrying away occurs, but coke-like layer is formed;

3) for flow temperatures between 394° and 617°C the material is destroyed without coke-like layer formation.

4) materials, operating at low heat flows with temperatures below 394°C; here the material is not destroyed and coke-like layer not formed;

5) a special case of operation in high heat flows - combination of low carrying-away of the material mass and low enthalpy change perpendicular to border layer, that induces exothermal mass transmission and surface chemical reactions.

In such cases no melt is observed at the cover surface. Exothermal conditions in the border layer result in surface temperature increase above the melting one. As a result the surface layer is packed or the surface is carried away at high aerodynamic shears. Though the melt layer at the surface is not observed, the surface carrying-away can be explained either by that of melt drops which could be formed between 1505 and 1721°C, as silicum oxide is the main product of thermal decomposition (melting temperature being 1617°C), or by the reaction $C+SiO_2 \rightarrow CO+SiO$ at temperatures above 1394°C.

ESM-1004 materials can be used in a wide range of entrance conditions with moderate heat. They can be reinforced by Nomex combs (the material ESM-1004NS for a stern part). For front parts one should use the material ESM-1004LP with the density 0.9 g/cm^3, for sides - ESM-1004AP with the density of 0.56 g/cm^3.

Covers for ESM-1004LPs were 25 mm thick at the braking point, and those from ESM-1004AP - 13.5 mm thick. The thickness of covers is defined from the SV entrance conditions. For tight entrance angles carrying-away of coke-like layer at the head surface appears most important, while for small trajectory incline isolation properties are paramount.

A number of USA companies try to create syntactic foams for heat-shielding purpose, including General Electric.

Such covers have densities above 560 kg/m^3, thanks to introduction of maximal quantity of quarts, glass or phenol-formaldehyde microspheres into the liquid elastomer. Materials of this type have the same ablation mechanism as rubbers based on silicon-organic elastomers. The materials of General Electric are rather more effective than those used for "Mercury" SV heat-shielding [70, 79-81].

Application of microspheres of one type in syntactic materials is considered inexpedient. Thus, when just glass microspheres are used, the density increases by approximately 60%, and when polymer ones - tensile rigidity decreases more than twice [82].

Martin Marietta Corporation has elaborated light HSM for SV designed for Mars [83-86]. Aerodynamic braking of SV in Mars atmosphere is rather high, the fact demanding effective heat shielding. The latter should be radio-transparent both before and after the ablation process. These demands can be satisfied either by a high-effective coking HSM with windows from radio-transparent material or less effective one, but radio-transparent itself. Cover efficiency being decreased somewhat, the total weight of heat shielding is increased. It does not prevent from its application, as radio-transparent windows in the high-effective cover also increase the construction weight because of edge fixtures.

There exists the necessity of elaborating "superlight" HSM with low heat conductivity and increased heat capacity as compared with known light materials, for instance, ESM-1004 of General Electric, phenol-nylon one with the density 400-600 kg/m^3, the one based on epoxy resin AVCO 5026-39, the materials based on silicon-organic resins NASA-62, ESA-3560 HF. It is explained by the fact that the Martian SV heat-shielding from these materials weighs more than its body. The conditions of SV entrance into the Mars atmosphere along two trajectories are shown in Table 10.

Table 10. The conditions of heat-shielding operation during SV entrance into the Mars atmosphere.

Characteristics	Trajectories	
	the first	the second
Atmosphere compositiona[I]	VM-7	VM-8
Time of gliding, s	250	50
[I]Atmosphere composition (model): M-7 - 28 weight % of CO_2+72 weight % of N_2;	3660	4880
M-8 - 100% CO_2.	-10	-18
Velocity, m/s		
Angle of SV entrance, grade	22700	13400
Maximal surface density of heatflow (in breaking points), kJ/m^2s		

One should note that for the considered trajectories the vessel heating is of convective character, the part of radiation heat flows now exceeding 5% from the total one.

The company has elaborated and tested from all points of view the materials of both types: SLA-561 of decreased efficiency and SLA-741 - high effective and coking.

Porous Teflon and compositions, forming not carbon, but silica solid sediment, were used as radio-transparent materials for antenna windows.

The properties of the materials were evaluated previously by the data of tests in the flame of propane burner at the heat flow 450 kJ/m^2s. Specimens were kept in the flame until temperature of their backside reached 246°. The test duration not exceeding 1.5 min, the temperature of SLA-561 specimen was increasing rather slow. it is explained by its good heat-isolating properties. After that the coke-like layer began to destroy and the temperature increased sharply. The temperature of the surface under the flame effect reached 1093° and changed no more.

Temperature distribution in the material layer can be defined by the formula:

$$T = T_o\left(1 - erf\frac{x}{2\sqrt{at}}\right),$$

where T - the temperature inside the specimen at the point x from the surface of heat supply;

T_o - the specimen initial temperature;

a - the coefficient of temperature conductivity;

t - the duration of flame effect;

erf - mistake function.

The formula was shown to be valid by the tests of specimens 7.2, 11.4 and 16.4 mm thick.

The tests of materials under the conditions initiating the entrance into the Martian atmosphere were performed to plasma-arc device with the power 200 kW.

The regimes of tests are shown in Table 11. Specimens (disks 64 mm in diameter) had various thicknesses to define the dependence of the back side temperature on the weight of the cover surface unit. The specimens were glued to a molten epoxy support and the temperature of their back sides registered by Chromel-Alumel thermocouples.

Table 11. The results of test at plasma-arc device of Martin Marietta company

Characteristics	Regime variants		
	1	2	3
Heat flow directed to a cold wall, kJ/m^2s	68	232	680
Plasma flow enthalpy, kJ/kg	6020	9500	15750
Braking pressure, atm.	0.0033	0.0059	0.0095
Test duration, s	250	150	50
Specimen surface temperature, °C	897	1129	1520

Notes: 1. Operating surrounding composition weight %: N_2 - 72, CO_2 - 28.

2. The first variation of the test regime is in accordance with the heat flow directed to a cold wall at the entrance speed 3660 m/s and angle 10°, the second one - 4580 m/s and the same angle, the third - 4980 m/s and angle 18°.

Relative test of various HSM were performed under these regimes. The most interesting results of these tests and thermal physical characteristics of the materials are shown in Table 11. It can be seen that SLA-561 forms the coke-layer much thinner than SLA-741.

Unlike phenol-nylon based specimen, SLA and cork ones are characterized by increasing thickness in the test process. The thickness of the coke-like layer (measured from the initial position of heat-feed to the border of fully coked layer) for SLA materials was about 6.35 mm for the first and the third test regimes and 8.1 mm for the second one. The pyrolysis zone thickness, defined as the distance from the coked layer to the border of unputrid material, for SLA materials was 5.0-6.35 mm for the first test regime and 0.76 - 1.78 mm for the third one.

Thermogravimetric analysis of these materials in argon atmosphere, the rate of temperature increase being about 3 grade/min, has shown that the decrease of the specimen weight becomes noticeable at 177°, at 370° its rate becomes maximal. Specimen decomposition is finished at 593°. Solid residues of SLA-561 and SLA-741 are 56 and 46%, respectively.

The atmosphere composition effect on HSM ablation was studied at plasma-arc device in nitrogen and carbon dioxide mixture in ratio 72:28 and 50:50 [84]. Test regimes were approximately the same. High reaction ability of the medium with carbon dioxide is explained by the fact that at high temperatures CO_2 is decomposed into CO and O_2. Oxygen concentrations in these experiments were 10 and 18 weight %. In SLA specimens there was no difference in weight loss, surface state and coking depth. It cannot be explained by the difference in gaseous medium composition. At the same time for phenol-nylon based control specimens with increased oxygen content a tendency was observed to weight loss increase.

The dependence of specimen back side temperature increase on the weight of cover surface unit shows the materials SLA and Armstrong-514 to be most effective. Despite the SLA-741 material provides the lowest temperature increase of the substrate, the SLA-561 was chosen as the final variant. Its coke-like layer has much lower cracking and shrinkage and is destroyed at heat flow values, twice as large.

Table 12. Thermal physical and heat-shielding properties of low density HSM.

Material	Density, kg/m^3	Coefficient of heat conductivity, W/m·grade	Specific heat capacity, kJ/kg·grade	Specimen thickness, mm	Temperature of specimen back side, ^oC			Coke layer thickness, mm			Carried-away layer thickness, mm			Surface state after tests
					Regime variations									
					1	2	3	1	2	3	1	2	3	
SLA-561	197	0.052	1.255	9.3	–	–	66	–	–	5.35	–	–	0.89	Separate splits
				12.2	104	72	–	5.6	7.87	–	0.25	0.74	–	The same
SLA-741	182	0.052	1.338	9.2	–	–	50	–	–	6.35	–	–	0.58	A lot of light splits
				15.0	53	33	–	6.86	8.39	–	0.58	0.36	–	Separate splits
Phenol-nylon based	520	0.108	1.586	7.6	–	–	17	–	–	3.05	–	–	0.76	The same
				13.0	24	28	–	3.56	4.32	–	1.01	1.52	–	Thin splits
Cork:														
Armstrong -514	303	0.055	2.090	5.8	–	–	47	–	–	3.56	–	–	0.94	Coke layer partly destroyed
				7.9	90	117	–	4.06	5.60	–	0.03	2.28	–	Separate splits
Material 2755	520	0.072	2.090	13.0	11	6	–	–	2.03	–	–	–	–	The same

Note: test regimes see in Table 11.

SLA-561 material is cold-resistant till -101°C, does not change during long keeping in vacuum at 71° and sterilization for 92 hours at 135° in nitrogen medium. The weight of 1 m^2 of heat-shielding cover providing the increase of the specimen back side temperature up to +111°C upon tests by regimes 1 and 3, is 2.34 and 1.61 kg respectively, i.e. depends sharply on the entrance conditions into the Martian atmosphere.

To clear possibilities to create heat-shield from the SLA-561 material for precise operating conditions, the heat transfer in ablation process was calculated by electronic computer. The data are in a good agreement with support experimental temperatures measured with the help of plasma-arc device.

To obtain these materials new ways of low-dense silicon-organic substance developing have been found [87]. According to one way high-effective coke-forming material is obtained, that contains not less than 92% of low-dense filler by volume, i.e. the ratio of this filler and other components of the material can reach 16:1. Moreover, the material could contain the bark of cork oak, silicon or glass microspheres, phenol microspheres and silicon or hydrocarbonic fibres. Another way gives the material, that forms solid silicon base at high temperature influence. It contains not less than 90% of low-dense filler by volume, and its ratio with other components can reach 11:1. Moreover, this material contains additionally silicon and glass microspheres and silicon fibers.

Both types of the materials are intended for external heat-shielding of SV, including challenging ones. As a result of contact with high-temperature border layers materials are affected by physical and chemical influence. Resulting this surface temperature is stated, and heat flow penetration in the material is sharply limited. The one of the abovementioned applications is for SV, entering Martian atmosphere at rate from 3.6 to 4500 m/s. Facing could be treated by heat influence during up to 250 s at heat flow of 1804 kW/m^2 and total heat influence of 17010 kW/m^2. Dynamic pressure during entrance will be low because of low density of Martian atmosphere. For these conditions other materials could be used, which are applied for more stiff conditions of the Earth atmosphere entrance. But they are too heavy for planetary ones, and if they were used, the weight of heat-shielding would be higher than the planet vessel one. That is why the problem of density decrease for planet vessels is the most sharp, because the problem of useful cargo is connected with it. Thus, for example, for 743-kg vessel the decrease of heat-shielding by 40% means the increase of useful loading for 45.3 kg.

Low-dense materials are also applied in HSM for challenging SV. However, their voluminal content is just 55-60%, that is 90% less than Martian Marietta Company suggests. However, their content by volume is just 55-60% or in comparison with these suggested by Martian Marietta Company by 90% lower. Special attention the Company pays to light filler, cork, also. HSM with cork are of density from 0.3 to 0.5 g/cm. Cork has lower heat conductivity and higher heat capacity in comparison with the basis of the material - elastomer, and corresponds to it higher ablation efficiency, respectively. However, cork is affected by abrupt shrinkage and splitting during ablation and, corresponding, forms structurally weak coke.

Table 13 shows properties of low dense coking HSM.

Table 13. Density, heat conductivity and heat capacity of low dense HSM.

Material	Density, g/cm^3	Heat conductivity coefficient, W/m•grade	Specific heat capacity, kJ/kg•grade
1	2	3	4
Phenol-nylon	0.52	0.86-1.31	1.59
ESA-3560 HF	0.51	1.01	1.22
NASA-602	0.58	1.27	1.68
DC 325	0.87	1.51	1.34
ESM 1004	0.62	1.78	1.38
Avco 5026 39-HC/g	0.50	0.88	1.81
Corkboard-Armstrong Insulcork 2755	0.53	0.58-0.86	2.10-2.52
Corkboard-Armstrong 514	0.30	0.55	2.10
SLA-561	0.225	0.52	1.26
SLA-741	0.215	0.52	1.34

Note: Test regimes see in Table 11.

In some cases for earth and planetary challenging SV it is quite desirable for heat-shielding cover to be radio-transparent both during and after ablation process for alleviation of radar signal penetration and for communication, also. Superthin HSM of Martin Marietta Company could be also applied for this purpose.

Table 14 shows features of different radio-transparent materials.

SLA-220 material is sufficiently lighter than teflon, melted quartz and boron nitride. Moreover, the latter three could be hardly connected with protected surfaces.

The dependence of material density on the among of microspheres installed is shown in the patent. Thus, if density is 0.3 g/cm^3 at 100 mass parts of microspheres installed, it equals 0.25 g/cm^3 at 150 mass parts, and 0.22 g/cm^3 at 200. If we use cork (without fibers) these values change as follows: at 50 mass parts density is 0.22 g/cm^3, at 60 it equals 0.20 g/cm^3 and 0.18 g/cm^3 at 80 mass parts.

Table 14. Density, heat conductivity and heat capacity of radio-transparent materials

Material	Density, g/c^{m3}	Heat conductivity coefficient, W/m•grade	Specific heat capacity, kJ/kg•grade
Teflon	2.15	2.45	1.05
Pored teflon	0.51	0.60	1.05
Pored teflon	0.71	0.63	1.05
Melted quartz	2.21	19.3	0.71
Borum nitride	2.27	328.3	0.80
SLA-220	0.25	0.79	0.97

The increase of microsphere content allows to decrease sufficiently the material density. However, the limit of tensile strength and specific elongation decrease also, as it is seen from Table 15.

SLA-220 material is characterized by high heat-shielding properties, capability to form film siliceous base on the surface. It is stable for sound influence up to 100 dB and IR-transparent [88]. Absorption coefficient is 0.16-0.21, irradiation coefficient equals 0.85-0.95 at temperature at 1371°C. It has been tested at maximal flow of $8*10^4$ W/m^2 with imitation of 100 launchings at temperature from 343 to 545°C. It also ran successfully through several cycles at 630°C during 120 s. During these tests the temperature of protected surface did not exceed 121°C. SLA-220 material was applied for the protection of some details of "Viking" SV. It was also used if hypersonic aircrafts of gliding type with the flight height from 15 to 30 km and speed up to 6M. The application of the material at temperature 815°C is not recommended [89].

Table 15. The influence of filler contention density, physical-mechanical properties and heat conductivity of syntactic HSM.

Material trademark	Density, g/cm^3	Filler content (mass parts for 100 mass parts of e elastomer)		Tensil strength limit, kgF/cm^2	Specific elongation, %	Heat conductivity at 195°, W/m• grade
		Low dense	Total filling			
ESA-5500	0.88	88	133	9.8	36.0	0.67
ESA-3560	0.48	153	155	10.5	9.5	0.29
SLA-220	0.25	1176	1182	5.3	6.2	0.25
SLA-561	0.225	1691	1702	4.3	2.5	0.15
SLA-741	0.215	1779	1790	3.7	3.4	0.15

The ratio by volume of filler and elastomer in SLA-561 material is 17.02:1, the ratio by weight is 3.06:1. The same ratios for SLA-741 and SLA-220 materials are 17.9:1 and 11.84:1 by volume and 3.06:1 and 2.23:1 by weight, respectively. Such high filling requires special technical approach. In the mixers used small shearing forces should exist for preventing of microsphere destruction. The pressure during cover formation should not exceed 1 atm. Viscosity of liquid silicon elastomer should be lower than 10000 cantipuaz.

To attach the ability of coke formation the following ingredients are introduced into the materials: cork, silicon and glasseous microspheres, phenol microspheres and silicon or carbon fibers.

The descriptions of SLA-561, SLA-741 and SLA-220 materials are shown in Table 16.

Table 16. Composition of heat-shielding materials SLA-561, SLA-741 and SLA-220.

Ingredients	**% by weight**	**% by volume**	**Volume parts**
1	**2**	**3**	**4**
SLA-561 material (0.225 g/cm^3)			
Elastomer (0.995 g/cm^3)	24.57	5.55	100.00
Silicon fibers (2.20 g/cm^3)	2.92	0.30	5.40
Carbon fibers	2.34	0.30	5.40
(1.85 g/cm^3)	35.10	43.90	790.99
Silicon microspheres			
(0.18 g/cm^3)	5.85	14.35	258.55
Phenol microspheres			
(0.092 g/cm^3)	29.22	35.60	641.51
Cork (0.186 g/cm^3)			
SLA-741 material (0.215 g/cm^3)			
Elastomer (0.995 g/cm^3)	24.57	5.30	100.00
Silicon fibers (2.20 g/cm^3)	2.92	0.28	5.28
Carbon fibers	2.34	0.28	5.28
(1.85 g/cm^3)	23.40	27.90	580.00
Silicon microspheres			
(0.18 g/cm^3)	5.85	13.64	257.35
Phenol microspheres			
(0.092 g/cm^3)	4.92	47.20	890.56
Cork (0.186 g/cm^3)	–		
Pority		5.40	101.88
	100.00	100.00	1890.35
SLA-220 material (0.250 g/cm^3)			
Elastomer (0.995 g/cm^3)[2]	31.0	7.8	100.0
Silicon fibers (2.20 g/cm^3)	4.7	0.5	6.41
Silicon microspheres	64.3	91.7	1175.64
(0.18 g/cm^3)			
	100.0	100.0	1282.05

The size of cork particles is from 0.4 to 0.75 mm, the sizes of phenol and silicon microspheres are from 0.07 to 0.2 mm., 2 Including vulcanizing system

As an elastomer one can use the industrially processed ones, for example: RTV-655 - methylphenylvynilpolysyloxane, RTV-602 - dimethylpolysyloxane and RTV-615 - dimethylpolysyloxane also, produced by General Electric Company. Phenol microspheres, filled with nitrogen, are produced and delivered by Union Carbaid Company, silicon ones - by Emerson and Caming Company, silicon fibers - by Thompson Fiber Glass Company, cartoon ones by Hitco Company, cork - by Armstrong Cork Company. Fillers could be treated by sylane - containing agent A-1100 Silane of Union Carbide Company, and by similar ones for improving connections and wettability with elastomer.

Materials are prepared by mixing elastomer, vulcanizing agent and fibers, treated by apprete, at low and middle rate of mixing during 2 min. After that they are mixing at high rate during 26 min. The last step is the adding of cork and microspheres, preliminarily mixed. Total mixing time is 60 min. Molding and vulcanization are performed in dismountable, for example, aluminum form at 50 kgF/cm^2 pressure and temperature of 121°C during 16 hours. The cover can be obtained also from three-component system (paste, vulcanizing agent and diluent) by deposition (evaporation).

As a material for microspheres one can use copolymers - from acrylinitrile and vinylidenchloride in heat-shielding covers for SV to prevent the influence of high-temperature turbulent gas flows (2600-3150°C [90]. Microspheres are filled by neopenthane. The compositions suggested could contain for 100 mass parts of elastomer from 0 to 250 mass parts of silicon-containing filler to increase tensile and shearing strengths. To increase the density from 0 to 15 mass parts of carbon fibers should be added in order to increase the adhesion connection on support-cover interface and from 1 to 50 mass parts of hollow microspheres.

Technology of cover producing is the following: the catalyst for elastomer vulcanization is added to the mixture heated up to 90-135°. Then the mixture is sprayed or smeared onto the articles of various forms, made from titanium, steel, etc.

To select heat-shielding cover for new aircraft model X-15-2A, with flight speed reaching 8800 km/hour, a great among of materials from different companies have been tested: General Electric Company, Martin Marietta Company, MacDonnel Aircraft Corporation, North American Aviation Company. Moreover, a number of materials were suggested by scientific research center from Langley [91,92]. At the height of 30 km metal surface of the aircraft without heat-shielding could be heated up to 1327°, metal heating being allowed just up to 650°. The materials with density from 433 to 1762 kg/m^3 were tested in hypersonic aerodynamic tube with electric-arc heating (gas flow enthalpy was 3490 kJ/kg, heat flow directed to cold side of a sample was 1430 kJ/m^{2}s, M=3.2). The best results were obtained for sprayed material MA-25S of Martin Marietta Company, which was based on silicon organic elastomer with density of 448kg/m^3. It is coking during the flight, and between flights coke-like layer is removed, and the initial thickness of the cover is restored by re-evaporation. MA-25S material was recommended for application at aircrafts with the flight speed over 6000 km/hour.

Special demands are put forward for heat-shielding, applied in challenging space systems, which represent piloting SV of various application and consist of 3 main parts. These parts are: pilotting orbital challenging unit, two solid-propellant missile accelerators and one-shot fuel unit [93-97].

Great problems appear in projecting of heat-shielding for fuel unit, that contains liquid hydrogen of -253°C temperature in one compartment and liquid oxygen of -183°C temperature - in the other one. So, heat-shielding should fulfill two tasks: shielding from high temperatures appeared during aerodynamic heating, on one hand, and from superlow temperatures, on the other hand. On the whole, this task is fulfilled by passive polyurethane foam heat-shielding. But in the zones, where heat influence is great, polyurethane foam is covered by active ablation heat-shielding from the external side.

Covering by heat-shielding is performed by the method of vacuum connection of preliminary produced panels with the help of X-6300 glue and DC-1200 undercoat. If fuel unit surfaces are of complicated geometry (nasal deflector, transverse bar), SLA-561 material is deposited by spray. After vulcanization the final mechanical processing is performed. Protection of polyurethane foam from the influence of dense heat flows is performed by removing the cover mass during ablation process. Carbonized layer is formed on the cover surface during heat influence. Gaseous products of thermochemical decomposition penetrate through its pores to cover-air interface and decrease convective heat exchange. Cover thickness at the nasal compartment of the oxidizer is 0.89 cm. It raises up to 2.54 cm at the transition of nasal compartment into cylindric shell and at the part of front cylindric shell. Then it decreases to 0.25 cm in the zone of back section of intermediate compartment, front section of fuel compartment and back linkage points of the second section to the fuel tank.

White waterproof cover SDC 92-007 is externally put on the silicon-organic cover.

Silicon-organic material ESA-5500 of higher density has been used for heat shielding of the front edges on the missile accelerator and of the front edges of space aircraft [93,94].

The application of silicon-organic heat-shielding compositions for repair of nonelastic heat-shielding covers is unusual [98-101]. Long functioning of SV at orbits is impossible without the control of devices and systems and their repair. The most important is the repair of heat-shielding covers at challenging transport SV.

SV were created with closed hermetic or open cabin, supplied with manipulators, brought to orbit in cargo compartment of transport SV and intended to perform repair works. There exists also a portable apparatus of Martin Marietta Company, tested by five space-flighters at "SkyLab" SV. It allows to perform repair works in space in a fixed state at the surface of plate heat-shielding cover. Portable device is equipped by power units of rectilinear and rotating movement with leverage and halyard for return if the latter one is out of duty. It is also equipped by mechanisms of repair composition injection to a damaged section and by repair kit with repair materials and equipment. The sizes of portable unit at 1250*815*660 mm, the mass is 136 kg.

The repair kit, worked out by this Company, is applied in Johnson's Space Center. It includes 160 profile plates of heat-shielding material of 150*150 mm size and 20-40 mm thickness. It also includes balloon with sprayer for injection of eight-component ablational HSM. Each component is contained in the balloon of 0.0025 m^3 volume (the total volume of one material portion is 0.0177 m3). Five sections of 0.915*0.457 m are selected on the surface of transport SV, where the loss of plates is mostly possible (three for each section). The repair kit is kept in cargo compartment of transport SV and is used by spacemen after controlling the status of the plate heat-shielding cover during orbital flight. The Company also worked out a repair panel for works at the orbit. It is equipped with four sprayers, eight devices for cover controlling, container for material storing at 27°C and 165 ceramic plates of various forms and sizes. using this panel, equipped with special instruments, one could repair splits up to 6 mm depth and change separate plates. The panel has been tested in imponderability on board the Boeing KC-135 aircraft.

The repair of heat-shielding covers can be performed by changing damaged plates, which are fixed on dempfing submarine at the main cover, by the plates from HSM. Heat-shielding properties of such plates were determined in conditions, imitated the landing of transport SV. The sample represents the ablational plate, surrounded by usual heat-shielding covers.

Tests show that if the repair plate is 3 mm jet out over usual ones hot gas flow penetrates into the split appeared then. This abruptly increases the temperature of the back side of the plate.

The following demands are set forth to the material vulcanized at the repair place: the temperature of transport SV main cover during landing should not exceed 177°C, vulcanization in vacuum at 4.5-52°C should take not more than 18 hours, working life not shorter than 1 hour, contact strength with contacting materials not less than 2.8 kgF/cm^2, storage period at 27°C not shorter than 6 months, density should be enough to guarantee not high thermal forces in surrounding plates. The weight of repair kit should not exceed 136 kg.

Two model syringe-mixers were developed and produced for preparation and injection of plate heat-shielding covers on damaged typical sections. These two are: independent one, in which mixing and material injection is performed by one device of nozzle type, and three part one, in which components are mixed in separate repair devices and repair mixture is injected into the syringe through the pipe to be put on the plate. Both these devices have pneumatic gear with gaseous carbonoxide injection with up o 14 kgF/mm^2 pressure. Dosed components are mixed with the help of three standard static heads of Ross and Son Company, in which the destroying aluminum membrane is foreseen. This membrane excludes contact of catalyst with the paste before mixing. Catalyst-paste ratio is regulated by concentric plungers; the total amount of mixture injected is 1180-1310 cm^3. Basing on the results of overland tests these devices should provide stable mixing of components and their feed to the repaired section, the possibility of one-hand work, simplicity of assembling and good visual range of repaired surface.

The repair materials have been worked out, which could be syringed and are fixed at damaged plates by RTV-560 hermetic of cold vulcanization, which is based on polysyloxane elastomer. To decrease density and provide necessary thermophysical properties silicon and phenol hollow microspheres or their combinations were introduced into polymer.

There are two materials created: RTV-560-55 with density of 0.88 g/cm^3, that is useful for processing in syringe-mixers and highly filled RTV-560-32 with diluent, useful for preparing repair plates in space conditions. Silicon microspheres were introduced into RTV-560-55, which form the melt on the surface of heat-shielding cover (5,6 mass parts for 100 mass parts of the paste), and cocking phenol ones (of the same amount). Into RTV-560-32 just silicon microspheres were introduced (33 mass parts for 100 mass parts of the paste). The composition of tin octoate and tin dibuyldilaurate in ratio 1:1 was used as the catalyst, which was mixed with microspheres and silicon oil. At the content of tin octoate of 1,25 mass parts per 100 mass parts of the paste working life and vulcanization rate of compositions fulfilled the demands.

Physical-mechanical and thermo-physical properties of vulcanizates were studied for the samples, prepared as follows. Microspheres were dried during over 16 hours at 60°C in vacuum thermostate, RTV-560 was degassed in vacuum. After that microspheres and the paste were mixed in vacuum and vulcanizing system fed on the air, water traces, necessary for vulcanization, leaving in catalyst mixture. The samples were vulcanized in vacuum and in the air. Table 17 shows the results of these tests.

Table 17. Properties of repair materials RTV-560-55 and RTV-560-32.

Properties (in vacuum)	RTV-560-55	RTV-560-32
Density, g/cm^3	0.73	0.48
Limit of tensile strength, kgF/cm2	14.2	10.6
Strength of fixing to RTV-560, kgF/cm2	4.36	6.17
Hardness by Shore "A"	70	66
Heat conductivity coefficient, W/cm•grade	$5.4\cdot10^{-4}$	$4.3\cdot10^{-4}$
Specific heat capacity, kJ/kg•grade	1.42	1.42
Temperature conductivity coefficient, m^2/s	$4.5\cdot10^{-8}$	$6.0\cdot10^{-8}$
Thermal coefficient of linear expansion, 1/grade	$1.8\cdot10^{-5}$	$1.8\cdot10^{-5}$

For determination of adhesion of RTV-560-55 and RTV-560-32 to the material of plate heat-shielding cover sample destruction followed the latter. Strength limit of the first material was 1.18 kgF/cm^2, and the second one was 0.65 kgF/cm^2.

Heat-shielding properties of repair materials were estimated at electric-arc device NASA. The samples under tests represented cylinders of 51 mm width and height, into which thermocouples were placed. The samples were prepared by folding on cylindric substrates made from plate heat-shielding cover of HRSI trademark in the air and in vacuum. The results of temperature determination of the samples of various thickness at their testing at arc device during 600 s are shown in Table 18.

Table 18. The results of tests of repair materials RTV-560-55 and RTV-560-32 at electric-arc device.

Material	Vulcanization	Thermocouple position depth, mm	Temperature, °C
RTV-560-55	in vacuum	0.0	1430
		6.6	899
		13.8	475
		26.3	–
	in the air	0.0	1367
		6.9	951
		13.7	428
		26.7	60
RTV-560-32	in vacuum	0.0	1444
		6.6	941
		13.0	433
		25.4	84
	in the air	0.0	1355
		6.4	980
		13.2	670
		25.4	142

The data shown in Table 18 state that both materials can provide heat-shielding of the main cover (temperature does not exceed 177°C) of transport SV if its initial temperature does not exceed 52°C. Coke separation from material was observed that could be eliminated by introducing heat-stable fibers into the material.

The analysis of the test results on this device allowed to determine that density, heat capacity and heat conductivity are of the second role at selecting repair materials. The possibility to overcome the influence of the flow treating transport SV at atmosphere entrance plays the main role.

Thus, elastic materials based polysyloxane elastomers with density of 0.5 g/cm^3 or a little bit higher are perspective from the point of technological (in space conditions) and heat-shielding properties.

One more application of compositions on the basis of silicon-organic elastomer for transport SV is their usage as one of the components of the space filler between ceramic plates in order to protect side surfaces of the plates and vessel main cover from overheating during entering dense atmosphere [102]. The filler consists of aluminoborosilicon tissue, impregnated by silicon-organic composition, which contains particles of an inorganic substance with high radiation ability. Resulting this, the filler also has radiation ability ≥0.7. Composition is vulcanized without heating. Dimethylpolysyloxane purified from ferric oxide was taken as the elastomer; tin dibutyldilaurate was taken as the catalyst; highly purified tetraboride silicide was taken as a filler with high radiation ability, its glass-fibers contain 7.25-7.75% of borum oxide.

RTV-560 compound with phenyl-containing elastomer and filled by ferric oxide can also be applied as silicon-organic composition. But this compound has decreased radiation ability and is powerful at temperature of 1315°C just for short time.

MATERIALS, REINFORCED BY CELL FRAMEWORKS

The materials, reinforced by cell frameworks, based on silicon-organics are used specially for external heat-shielding of aircrafts. Most interesting are the materials, which represent combination of cell framework and easily frothing or syntactic filler. Such materials are particularly efficient at shearing forces under the influence of air flow, which exceed 0.004 kgF/cm^2, and heat flows over 3490 kJ/m^2s [72-74, 83].

Frothing materials of ESM-1000 type by General Electric Company are used as fillers of phenol-fiber glass cells with cut cells of 6.2 mm size. Cell framework of the material is grounded to provide good linkage with fluid filler. The latter is prepared by thorough mixing of elastomer with fillers, vulcanized agents and are molded in the air to the plate of the required size. After that the framework is installed into the liquid filler until it touches the bottom of the form. Depending on the type and dosing of the catalyst and frothing, the material is stored several hours at room or at increased (50-60°) temperature to complete the first stage of vulcanization. After that the material is removed from the form and is vulcanized up to completeness at 150°C during 10 hours.

For heat-shielding of "Jeminy" space vessel Mac Donnel Aircraft Corporation used reinforced DC-325 material of Dow Corning Company [103-109]. It is intended for application at 3320-8400°C temperatures. It is of the density twice lower than fiber glass plastics and is rather high elastic, has low heat conductivity and high specific heat capacity. It also has rather high value of irradiation coefficient in infrared band and absorption coefficient in the Sun spectrum. This material is resistant to the influence of oils, fuels, hydrosystem liquids, dissolvents, hydrogen peroxide, nitrogen trioxide and assymetric dimethylhydrasine [110].

Preliminary studies of heat-shielding properties of the material were performed with the help of oxygen-acetylene burner at heat flows from 128 to 11300 kJ/m^{2}s and in high-rate air flow at 2760°C at 3M rate and heat flow of 11280 kJ/m^{2}s [ ]. At low heat flows DC-325 material excels other polymer HSM of that type, and at high heat flow does not reside in them. At the lowest heat flow stable coke-line base of 0.4 g/cm^3 is formed at the cover surface. The influence of high temperatures causes the formation of viscous glassy-like film, that decreases carrying away of cover mass in high-rate gas flow. Solid base formed under this film is of maximal strength at the highest heat flows. A very thin layer of the material, partly decomposed by heat, exists under the solid base. The rest of its mass is unchanged. It has been determined that high pressure and low enthalpy of gas flow causes the swelling of material surface layer. But at low pressure and high enthalpy of the flow the cover thickness decreases. It is observed that the formation of stable coke-like base is the result of complex interaction between elastomer and fillers.

The main properties of DS-325 material filler in vulcanized state are shown below:

Voluminal mass, g/cm^3	0.87
Solidity by Shore (scale A)	68
Lengthening of rupture, %	100
Tensile strength limit, kgF/cm^2	22
Modulus of elasticity at shrinkage, kgF/cm^2	154
Friability temperature, °C	68
Strength at uniform removal from aluminum substrate, kgF/cm^2	24.5

In nonvulcanized state the filler represents a paste of white color of 1500 ps viscosity, which is vulcanized at room temperature during 12-24 hours. The same process at 65° takes 4 hours and 15 minutes at 150°C. During preparation of heat-shielding for "Jeminy" space vessel the paste is poured out into the form of 2.75 m diameter, in which the framework from fiber glass fabric cells is placed. Obligatory condition of nondefect cover obtaining is the purity of the apartment, that should be nearly sterile [112]. Before fixing the cover on the protected surface the control by γ-radiation defectometer is performed in order to provide full absence of bubbles of air or admixtures. Cover sections which have defects can be easily removed and changed by new ones, for example, by infliction of the paste by putty knife with the following trimming of the surface.

At the tests at plasma-arc device carrying away of the material mass decreases sufficiently at introduction of phenol-fiber glass cell framework into superlight HSM of Martin Marietta Company [84, 85]. The test results for SLA-561 and SLA-741 materials (reinforced and reinforced by cells) are shown in Table 19. It follows from this table that introduction of cells increases the density of the materials by 11-12%. It, as a rule, twicely decreases mass carrying away of SLA-561 at the tests by 2 and 3 regimes, and by three times of SLA-741 material at the tests by regimes 1 and 2. The thickness of coke-like layer does not change practically in all the cases (test regimes are shown in Table 11). Comparing the back sides of samples of similar thickness one can see that heat transition increases a little in the presence of cell framework. As a result the specific weight of the square unit of the reinforced cover that provides similar heating of the protected surface will be 15-22% higher than that of the irreinforced one.

Boeing Company performed tests of three Martin Marietta Company's materials: SLA-561, ESA-3560 and ESA-5500. These materials were reinforced by fiber glass cells and were tested as heat-shielding for space tug-boat for different trajectories of the flight [113]. Low-temperature elasticity, simplicity of preparation and stability in the flow (heat flows in dependence on the flight trajectory were changing from 386 up to 1437 kW/m^2) were taken into account while selecting the materials.

ESA-5500 material of the highest density was selected for the conditions of maximal heat flow. So, respectively: the lower heat flow is, the lower dense material was selected. The thickness of covers was selected according to the results of high-temperature tests. It was found from 48.8 up to 54.1 mm. Cover weight in these conditions was from 11.4 up to 46.4 kg/m^2. Projectory mass of the heat-shielding at the entering dense atmosphere must be 323 kg.

Table 20 shows the properties of the materials selected.

Table 19. The results of testing irreinforced and reinforced materials of SLA type.

Material	Average density, g/cm^3	Sample thickness, mm	Back side temperature, °C			Coke thickness, mm			Removed layer thickness, mm		
			Variants of the regimes								
			1	2	3	1	2	3	1	2	3
SLA 561	0.227	7.6	-	-	139	-	-	5.85	-	-	0.46
with		12.1	-	92	-	-	5.85	-	-	0.43	-
cells		14.5	64	-	-	5.60	-	-	0.30	-	-
irrein-	0.197	9.3	-	-	66	-	-	5.35	-	-	0.89
forced		12.2	104	72	-	5.60	7.87	-	0.25	0.74	-
SLA-741 with cells	0.21	9.0	-	-	66	-	-	6.85	-	-	0.61
irrein-	0.18	9.2	83	94	-	6.35	8.13	-	0.20	0.23	-
forced		12.1	-	-	50	-	-	6.35	-	-	0.58
		15.0	53	33	-	6.86	8.39	-	0.58	0.63	-

Table 20. Properties of heat-shielding materials of space tug-boat.

Properties	Materials		
1	2	3	4
Density, g/cm^3	0.24	0.49	0.89
Coke density, g/cm^3	0.135	0.17	0.50
Specific heat capacity, J/kg•grade (at 21°C)	1.25	1.09	0.79
Specific coke heat capacity L/kg•grade (at 535°C)	1.55	1.22	1.55
	1.64	8.65	16.40
Heat conductivity coefficient, W/m•grade•10^6 (at 1370°C)	1.21	2.25	5.19
Heat conductivity coefficient, W/m•grade•10^6 (at 150°C)	2.52	2.52	2.52
	2	2	2
Specific heat of gaseous product formation, kJ/kg °C	0.44	0.65	0.44
	4.2	10.5	9.8
Degradation reaction degree	10.5	24.6	28.1
The part of gaseous products	2.5	9.5	36.0
Tensile strength limit, kgF/cm^2 at 21°C	3.1	17.0	60.0
at -85°C			
Comparative lengthening at 21°C			
at -85°C			

ESA-3560 material of brown color represents filled silicon-organic composition without cells or with cells from phenol fiber glass. It has density twice higher than the material of SLA type [96, 114]. It is freeze stable (up to 101°C), and it is well coking (37% base). The coke removing is negligibly small at temperatures up to 1650°C. But it becomes noticeable at higher temperatures.

The possibility of this material application as the heat-shielding of the front borders of challenging space aircraft, entering dense atmosphere layers with the use of air dynamic braking, is mentioned in the prospect of Martin Marietta Company [115]. It is pointed out that ESA-3560 cover in conditions of aircraft raising is treated up to the depth of 4.3 mm and is swelled up a little (0.8 mm). That is why the roughness appears at the surface. To preserve the surface aerodynamically smooth surface after entering the orbit and to prevent destruction at dense atmosphere layer entrance, the additional protective layer is put on it. This layer contains sorbing salts, which decompose with endothermal effect in the temperature range of 145-300°C.

ESA-5500 material is of black color and relatively high density (0.88 g/cm^3). It has higher heat conductivity than ESA-3560 material. However, it is destroyed with lower rate with the output of solid base of 57% and is more stable for erosion at temperatures over 1650°C [116]. Thus, if ESA-3560 material is fully removed after the double influence of the flow, that is 5.1 kW/m^2 power and 1630 s total duration, than ESA-5500 material is just swelled. The latter was successfully tested at the nature models of the Space Shuttle for the protection of the front borders of the rudders. The application recommended for heat-shielding material is the following: ESA-5500 is better for application as the construction material for protection at heat flows with the power over 114 kW/m^2, ESA-3560 is better applied at heat flows from 28 up to 114 kW/m^2, and SLA-561 is the best at the flows lower than 28 kW/m^2 [113].

CONCLUSIONS

1. HSM on the basis of silicon-organic elastomers do not split at deformations of protecting surfaces, heat and mechanic blows and vibrations, during storage, transportation and flight of missiles and SV. They show good heat-shielding properties and erosion strength, high freeze-, heat-, atmosphere- and humidity stability, the stability for degradation. They are also resistant to the influence of radiation and deep vacuum.

2. Silicon-organic HSM are prepared as monolithic or frothing rubbers or as the panels of low density (chemically frothing or made with the help of hollow microspheres - syntactic foams). They can be easily reinforced by fibers, tissues or cells. HSM infliction on the protecting surfaces is performed by gluing nonvulcanized, partly or fully vulcanized plates, or molded semifabrics, layer by layer infliction with the following mechanical processing, frothing without heating in open forms, spray, etc.

3. Silicon-organic HSM are rather widely applied to protect surfaces of super- and hypersonic aircrafts from aerodynamic heating and products of missile fuel burning, launching during the flight, and the external surfaces of missiles and piloted SV, either.

REFERENCES

1. *Missiles and Rockets*, v. 6, No. 19, p. 21 (1960).
2. C. Wipple, J. Thorme. Performance of elastomeric silicones in ablative and space environments. *Rubber Chemistry and Technology*, v. 39, No. 4, p. 1247-1257 (1966).
3. *Industrial and Engineering Chemistry*, v. 53, No. 8, p. 28A (1961).
4. *Aerospace Management*, v. 4, No. 11, p. 31 (1961).
5. *Materials in Design Engineering*, v. 56, No. 4, p. 81 (1962).
6. *Machine Design*, v. 33, No. 16, p. 14 (1961).
7. *Missiles and Rockets*, v. 9, No. 1, p. 21 (1961).
8. *Materials in Design Engineering*, v. 54, No. 2, p. 113 (1961).
9. *Missiles and Rockets*, v. 7, No. 18, p. 47 (1960).
10. *Aviation Week and Space Technology*, v. 85, No. 20, p. 114 (1966).
11. *Technology Week*, v. 19, No. 21, p. 4 (1966).
12. *Space/Aeronautics*, v. 47, No. 5, p. 129 (1967).
13. *Aviation Week and Space Technology*, v. 86, No. 1, p. 51 (1967).
14. *Rubber World*, v. 152, No. 5, p. 69 (1965).
15. *Materials in Design Engineering*, v. 63, No. 4, p. 156, 162 (1966).

16. *Aviation Week and Space Technology*, v. 76, No. 10, p. 100 (1962).

17. RTV Silicon Rubber, General Electric, Silicone Products Department, Waterford, N.Y. 12188. Company information leaflet.

18. Silicones. RTV Silicone Rubber, General Electric, Silicone Products Department, Waterford, N.Y. Company information leaflet.

19. *Chemical Engineering*, v. 72, No. 22, p. 110 (1965).

20. *Mechanical Engineering*, v. 89, No. 10, p. 57 (1967).

21. *The Engineer*, v. 257, No. 6643, p. 34 (1983).

22. *Rubber Age*, v. 98, No. 7, p. 130 (1966).

23. *Space Age News*, v. 10, No. 8, p. 21 (1967).

24. R. Anderson. New design concepts and materials and vigorous research, are openity avenues for reducing the weight of critical aerospace structures. *Astronautics and Aeronautics*, v. 2, No. 12, p. 14-20, 30-37 (1964).

25. W. Bobear, I. Thomas. Thermal Insulative Features of Silicone Rubber. VII *National SAMPE Symposium. Adhesives and Elastomers for environmental Extremes*, Preprint (1964).

26. *Rubber World*, v. 150, No. 4, p. 35 (1964).

27. *Rubber World*, v. 154, No. 1, p. 126 (1966).

28. *Design News*, v. 21, No. 5, p. 127 (1966).

29. *Aviation Week and Space Technology*, v. 84, No. 16, p. 13 (1966).

30. Patent USA No. 3506607, (1970).

31. C. Whipple, Silicones for a variety of aerospace applications. *VI National SAMPE Symposium Society of Aerospace Material and Process Engineering*, Preprint (1963).

32. M. Baile, S. Fuson. *Sealing the Future. Aerospace Composites and Materials*, v. 2, No. 1, p. 14-17 (1990).

33. Patent France No. 13773192 (1964).

34. D. Overmier. Development of Airframe Heat Protection for High-Performance Sounding Rockets. *Journal of Spacecraft and Rockets*. v. 4, No. 11, p. 1500-1504 (1967).

35. *Aviation Week and Space Technology*, v. 83, No. 1 p. 36, 37, 39 (1965).

36. *Aviation Week and Space Technology*, v. 87, No. 4, p. 75 (1967).

37. *Design News*, v. 23, No. 13, p. 16 (1964).

38. *Missiles and Rockets*, v. 14, No. 3, p. 23 (1964).

39. *Materials in Design Engineering*, v. 58, No. 6, p. 169, 170 (1963).

40. *Iron Age*, v. 193, No. 11, p. 169 (1964).

41. *Aviation Week and Space Technology*, v. 89, No. 8, p. 76, 77 (1968).

42. *Aviation Week and Space Technology*, v. 38, No. 26, p. 66 (1968).

43. Patent USA No. 4581391 (1986).

44. *Materials Engineering*, v. 95, No. 6, p. 24 (1982).

45. *Product Engineering*, v. 95, No. 4, p. 91 (1966).

46. *Materials Engineering*, v. 95, No. 6, p. 24 (1982).

47. S. Fixler. Armament Exhaust Structural Heating Effects on the F-14 Aircraft. *AIAA Paper*, No. 76-445, p. 10 (1976).

48. *Patent FRG* No. 1771772 (1977).

49. *Patent USA* No. 3553002 (1971).

50. C. Whipple Elastomeric Silicones for Aerospace Insulation. Section I. Ablative Materials. *8th National SAMPE Symposium: Aerospace-Hydrospace*, San Francisco, California, Preprints (1965).

51. G. Wehmueller. Elastomeric silicones for aerospace insulation. Section II. Heat recoverable materials, *8th National SAMPE Symposium: Aerospace-Hydrospace*, San Francisco, California, Preprints (1965).

52. J. Mellin, H. Sparhaw. Elastomeric materials for Venus entry heat protection. *AIAA Paper*, No. 201, p. 10 (1970).

53. *Product Engineering*, v. 37, No. 4, p. 91 (1968).

54. *Chemical Engineering*, v. 72, No. 15, p. 168 (1965).

55. *Missiles and Rockets*, v. 16, No. 20, p. 55-56 (1965).

56. *Plastics Technology*, v. 11, No. 3, p. 27 (1965).

57. *Materials in Design Engineering*, v. 57, No. 2, p. 7 (1963).

58. *Chemical Engineering*, No. 23, p. 115, 117 (1963).

59. J. Judge. New silicone rubber foams detailed. *Missiles and Rockets*, v. 16, No. 22, p. 81, 82 (1965).

60. *Aircraft Engineering*, v. 11, No. 37, p. 354 (1965).

61. *Industrial Finishing*, v. 41, No. 9, p. 4 (1965).

62. *Electronics*, v. 38, No. 21, p. 91 (1965).

63. *Rubber Age*, v. 97, No. 5, p. 69 (1965).

64. *Missiles and Rockets*, v. 17, No. 15, p. 40 (1965).

65. *Rubber World*, v. 152, No. 3, p. 101 (1965).

66. *Rubber and Plastic Age*, v. 46, No. 10, p. 1122 (1965).

67. *Plastic Week*, v. 25, No. 23, p. 1 (1965).

68. *Engineering Materials and Design*, v. 8, No. 11, p. 853 (1965).

69. C. Dolan, H. Edigheffer, P. Cline, E. Rienikka. Elastomer thermal shield systems for lifting re-entry vehicles. *AIAA Bulletin*, v. 2, No. 3, p. 93-96 (1965).

70. C. Dolan, H. Edighoffer, P. Cline, E. Reinikka. Elastomeric thermal shield systems for lifting re-entry vehicles. *Journal of Spacecraft and Rockets*, v. 3, No. 6, p. 867-872 (1966).

71. *Missiles and Rockets*, v. 16, No. 7, p. 40 (1965).

72. J. Leone, C. Maag. Thermal Radiative Characteristics of Low Density Charring Ablators for Planetrary Studies. *AIAA Paper*, No. 16, p. 1-6 (1969).

73. *Journal of Spacecraft and Rockets*, v. 5, No. 11, p. 1278-1284 (1968).

74. *Electronic News*, v. 13, No. 636, p. 51 (1968).

75. D. Florence, H. Thibqult, A. Hiltz. Selection, development, characterization and flight test of a thermal protection system for an Earth Entry Satellite Vehicle. *AIAA Paper*, No. 861, p. 1-5 (1978).

76. A. Chapman, M. Dow, Arc-Tunnel evaluation of some ablative heat shield materials for the x-15-2. Langly Research Center Langly Station, Hampton. National aeronautics and space administration. *NASA TND-3753*. Preprint.

77. J. Quillinan. A three-man space escape system. *Journal of Spacecraft and Rockets*, v. 6, No. 11, p. 1229-1236 (1969).

78. *Mechanical Engineering*, v. 89, No. 10, p. 57 (1967).

79. *Aviation Week and Space Technology*, v. 80, No. 21, p. 26 ((1964).

80. *Space/Aeronautics*, v. 41, No. 3, p. 15 (1965).

81. *Product Engineering*, v. 37, No. 13, p. 44 (1966).

82. *Patent USA* No. 358, 5157 (1971).

83. *Journal of Spacecraft and Rockets*, v. 3, No. 6, p. 867-872 (1966).

84. E. Strauss. New ablative heat shield materials for Mars Landers. *SAE Preprint.*, S.A., No. 660654, p. 1-12 (1968).

85. E. Strauss. Performance of superlight ablators is simulated Mars entry, including micrometeoroid damage. *Journal of Spacecraft and Rockets*, v. 5, No. 11, p. 1357-1360 (1968).

86. M. Jaybalan. A Study of Polymer foam blends for external application of space vehicles. *Elastomery and Plastics*, v. 16, No. 3, p. 175-178 (1984).

87. *Patent USA* No. 4031059 (1977).

88. Cited by Express-information VINITI. *Thermostable plastics*, No. 19, Paper 124, p. 14.

89. *Ablative heat shield materials Martin Marietta USA company information* leaflet (1982).

90. *Patent USA* No. 3734768 (1973).

91. *Interavia Air Letter*, No. 5920, p. 8 (1966).

92. A.I. Chapman, M.B. Dow. Arc-tunnel evaluation of some Ablative Heat Shield Materials for the x-15-2, *NASA TND-3753*, Preprint (1967).

93. G. Daforno, J. Graham, S. Tompkins. Initial development of an ablative leading edge for the space shuttle orbiter, AIAA 8th Thermophysics Conference, July 16-18, 1973. *AIAA Paper*, No. 739, p. 1-11 (1973).

94. *Aviation Week and Space Technology*, v. 94, No. 23, p. 55-61 (1971).

95. *Industrial heat*, v. 46, No. 8, p. 40-41 (1979).

96. E. Strauss. Ablative overlays for space shuttle leading edge accept heat protection. *AIAA Paper*, No. 737, p. 1-6 (1975).

97. D. Don, R. Mueller, K. Carnaha. Ablative materials evaluation for high performance lifting entry vehicles. *Space simulation processing symposium*, New York, Preprint (1972).

98. B. Smith, Backpack modified for tile repair use. *Aviation Week and Space Technology*, v. 113, No. 7, p. 65, 67, 68 (1980).

99. H. Hohler. Lusatzline arbeit for astronaution des space Shuttle, VD1, *Nachrichtendienst*, No. 15, s. 35 (1981).

100. S. Tompkins, C. Pittman, A. Stacey. Thermal performance of a mechanically attached ablator tile for on-orbit repair of Shuttle TPS. *Science and aerospace reports*, v. 18, No. 17, p. 2234 (1980).

101. C. Blevins, R. Gonzales. Development of a cure-in place ablator for space shuttle insulation. *JAA*, v. 21, No. 21, cat. 27, p. 3656 (1981).

102. *Patent USA* No. 4308309 (1981).

103. *Missiles and Rockets*, v. 14, No. 14, p. 19 (1964).

104. *Ceramic Industry*, v. 85, No. 4, p. 50 (1965).

105. *Materials in Design Engineering*, v. 58, No. 2, p. 107 (1964).

106. *Materials in Design Engineering*, v. 59, No. 2, p. 8 (1964).
107. *Machine Design*, v. 35, No. 15, p. 12 (1963).
108. *Material in Design Engineering*, v. 61, No. 3, p. 118 (1965).
109. *Aerospace Materials from Dow Corning*. Dow Corning 325, Engineering Products Division, Midland, Michigan.
110. C. Whipple, Piper, A. Smith, Silicones for ablative applications. *VII National SAMPE Symposium Adhesives and Elastomer for Environmental Extremes*. preprint. (1964).
111. *Iron Age*, v. 193, No. 13, p. 95 (1964).
112. *Plastics World*, v. 23, No. 4, p. 67 (1965).
113. E. Strauss. Ablative Thermal Protection for Space Tug Multipass Aerobraking Entry. *Journal of Spacecraft and Rockets*, v. 12, No. 6, p. 346-350 (1975).
114. *Aerospace Daily*, v. 106, No. 5, p. 36, 37 (1980).
115. Cited by Translation of TSNTI "Volna" No. 2711 *Ablative heat shield materials*. Martin Marietta Company information leaflet (1983).
116. *Aviation Week and Space Technology*, v. 108, No. 12, p. 60 (1978).

Elastomeric Heat-Shielding Materials for Internal Surfaces of Missile Engines

A.A. Donskoy
All Russian Institute of Aviation Materials
17 Radio Str., Moscow, Russia

Materials for shielding the walls of missile engines of solid propellant (MESP) are influenced by restoring gas flow of propellant combustion products under the pressure, reaching 150 and over atmosphere, and temperature up to 4000 K and the rate of tens and hundreds meters per second. At the start in solid propellant direct flow missile engine (DFME) conditions of the influence are similar to the above described ones, and on march the medium changes to oxidative one at the increase of the flow rate.

Thus, heat shielding materials (HSM) of ablation type for MESP and DFME must work at temperatures, which a fortiori exceed these of their thermal decomposition, and in conditions of transition from one physical state to another. Moreover, HSM must be technologic, light, atmosphere stable, vibro- and blow-resistant, humide stable and freeze strong. They must possess low thermo- and temperature conductivity, low thermal expansion coefficient.

In connection with the increase of sizes and the decrease of engine covers rigidity HSM must be elastic also in order not to be destroyed at deformations of shielding constructions under the influence of mechanical and heat stresses.

The most perspective direction of HSM processing is the application of elastomers, which allow to produce rubbers, foamy rubbers and combined rubber-fabrique or rubber-fibrillar materials. However, elastomers are linear polymers, that is why the most of them form no solid carbonic residue at thermal decomposition, i.e. they possess no coking ability without introducing special additives.

For the first time the problem of elastic HSM processing for put forward abroad in early 60s [1]. At present time sufficient information is stored about their properties and application. However, in some cases the information about the recepture of elastomer based compositions is limited.

HSM for internal surfaces of MESP and DFME must possess satisfactory freezing stability, high mechanical strength, coking capacity, high and stable adhesion to protective surfaces and to propellant also, if applied for strongly linked solid-propellant grains [2,3]. During engine performance shielded surfaces must not be heated over 150°C, mass and thickness of the coating should be minimum [4]. It is pointed out, that there are rubbers, worked out in USA, which allow to decrease weight of heat shielding of the engine twicely, approximately, and to use propellant, possessing higher temperature of combustion, than that previously applied. High-temperature tests allowed to obtain, that rubber like materials are the best for heat-shielding of MESP vessels. Such materials perform the formation of poured coke-like layer, possessing low heat conductivity, on the surface of covers under the influence of restoring gas flow. At thermal decomposition of the lower layers of the cover volatile products are extracted, which pass through the "coke" and create border gas layer, playing the role of additional heat resistance. According to the author's opinion, ideal heat shielding is the one, which is completely decomposed at the end of propellant combustion.

Table 1 shows the data on the main components, participating in the composition of some rubbers and rubber-like materials, applied in USA for MESP heat shielding, and their physicomechanical and thermophysical properties, also [5].

As it is seen from the Table, these materials possess very high elasticity (elongation at break is found in the range of 269-830 %, excluding M808 material, filled by asbestos), and sufficient mechanical strength, which reaches 266 kG/cm^2.

Various elastomers are used as the basis of HSM in many other states, also, for example, France and Australia. Thus, French Patent [6] mentions full number of the materials, based on polar rubbers, such as copolymer of ethylene and vinylacetate, containing 35-60% of the latter, butadien-nitril rubbers, able to contain carboxile groups, for example, copolymers of butadien, acrylonitril and acrylic acid, polychloroprene. Highly active silicon dioxide and (or) mica-like material is used as the filler. Layered filler content can reach 30 weight %, and 120 elastomer weight % - with silicon dioxide. Rubbers are vulcanized by organic peroxides or metal oxides. Adhesives are used for the fixation of vulcanized plates on metal. There is an example of the applied mixture: ethylene-vinylacetate copolymer (100%), silicon dioxide (40%), 40% solution of dicumyle peroxide

(4%), dispergator - stearic acid (1%), antiaging substance 0.5 phenyl-α-naphtylamine (0.5%), vulcanization activator - trialkylcyanurate (2%), antioxidant 2-polycarbodiimide (2%).

Table 1. Composition and properties of heat shielding rubbers

No.	Material mark	Caoutchouc	Fillers	Density, g/cm^3	Strength limit at break, kGs/cm^2 (at 21°C)	Elongation at break, % (at 21°C)
1	2	3	4	5	6	7
1	GTR-V-	Butadien-nitril	Silicon	1.28	77-105	300-450
	44	(Buna N or NBR)	dioxide,			
			asbestos			
		–"–	Silicon	1.22	175	400
	GTR-V-		dioxide			
	45	–"–	Silicon	1.25	203	610
			dioxide			
2	Garlock	Butadien-	Carbon	1.07	110	683
	7765	styrene (Buna S or	black,			
	GTR-V-	SBR)	asbestos			
	62		graphite			
			Graphite,	1.17	266	700
		–"–	carbon			
			black,			
	Goodyea		phenil			
	r		resin,			
	M800	–"–	silicon	1.39	42-98	60
3			dioxide	1.39	40	269
		Butylcaoutchou	Asbestos			
	Goodyea	c	Silicon	1.08	–	–
4	r M808		dioxide			
	AGC-		Silicon			
	MMB4	Polysulfidic	dioxide,			
			potassium	1.23	84	400
5			titanate			
	EP-2005		Asbestos,			
		Polyisoprene	silicon	1.19	105	–
6			dioxide			
			Phenolic			
	HITCO-	Polypropylene	microsphe			
7	2800		res,	1.41	52	350
			carbon			
	PFC-11	Ethylene-	black,			
		propylene	graphite			
8			Silicon	1.19	56	830
	SMR-81-		dioxide,			
	15		potassium			
		Polyurethane	titanate			
			Asbestos			
	P531 A					

Hardness by Shore (Scale A)	Heat conductivity coefficient, W/m* grad (at 93°C)	Specific heat capacity, kJ/kg• grad (at 65°C)
8	9	10
70	0.224	1.840
63	0.218	1.715
83	0.248	1.882
70	0.170	1.630
58	0.172	1.674
80	0.434	1.882
65	0.242	1.674
–	0.202	1.462
85	–	–
–	0.105	–
–	0.202	1.674
66	–	–

Note: Carbonic and graphite fibers may be introduced into the material composition, based on caoutchoucs, mentioned in No. 4 - 8, in order to increase their erosion stability.

Such materials possess enough thermostability, good propellant compatibility.

In Australia HSM, based on rubber linkage with ammonium perchlorate on the basis of nitrocellulose and nitroglycerin, for MESP with solid-propellant grains were worked out by the Department of Armament Scientific Investigating Institute of Procurement Ministry [7]. The materials are applied in the following conditions: propellant combustion temperature 1370-2760°C, pressure in combustion chamber 35 - 140 kG/cm^2. Combustion chamber cover is usually prepared for steel, sometimes from reinforced plastic; wall thickness lies in the range of 1-2.5 mm. Heat shielding coating is prepared from nonvulcanized sheets of rubber mixture, adhesed by special substances to the shielded surface with the following vulcanization of rubbers in combustion chamber.

Rubber mixtures are prepared on the basis of natural as well as synthetic caoutchoucs - chlorinated polyethylene CPE500 of Allied Chemical Company, chlorine sulfonated polyethylene Highpalone 20 of Du Pont Company, butadien-styrene rubber SBR 1502, 1712EP, 1778 of Austrapol Company, ethylene-propylene rubber Royalene 301 of Uniroyal Company, polybutadien 1220 of Austrapol Company.

The main applied fillers are also highly dispersed silicon dioxide and chrizotilic asbestos of 5RO4 mark. In this case filling degree must be as high as possible (about 50% usually). Combination of these two fillers is the most efficient from the point of view of heat shielding properties. The introduction of small amounts of fillers and the type of vulcanizing system does not practically influence HSM efficiency. It is pointed out, that only the content of silicon dioxide and asbestos has been changed during investigations of materials, based on various rubbers.

Ablation indexes were obtained in accordance with ASTM No. E285-65T method on fire of oxygen-acethylene burner. Standard material DW-7942, based on chlorine sulfonated polyethylene, was used for the comparison. Best results were obtained at testing materials, based on chlorinated and chlorine sulfonated polyethylenes and ethylene-propylene rubber. Testings of various materials showed also, that even the worst of them - the one, based on polybutadiene - possesses higher heat shielding properties, than standard material DW-7942.

Table 2 shows the content and heat shielding properties of Australian HSM.

Table 2. Composition and the results of high temperature tests of heat shielding rubbers

Ingredient	230/5	232/15	234/21	235/26	236/31	241/40	DW 7942
1	2	3	4	5	6	7	8
Chlorinated polyethylene CPE500	100	–	–	–	–	–	–
Butadien-styrene caoutchouc SBR 1772EP	–	100	–	–	–	–	–
Natural rubber	–	–	100	–	–	–	–
Ethylene-propylene rubber Royalene 301	–	–	–	100	–	–	–
Butadien-styrene caoutchouc Austrapol 1200	–	–	–	–	100	–	–
Chlorsulfonated polyethylene	–	–	–	–	–	100	100
High dispersed silicon dioxide Ultrasil VN3	44.5	44.5	44.5	44.5	44.5	44.5	Components are unknown
Asbestos of 5RC4 mark	65	65	55	55	55	55	-"-
Polyethylene AC617	1	–	–	–	–	4	-"-
Resin Stabelite	1	–	–	–	–	2.5	-"-
Magnesium oxide	5	–	–	–	–	5	-"-
Stearic acid	0.5	2	1	1	–	–	-"-
Chlorinated paraffin Cereclor 42	7.5	–	–	–	–	–	-"-
Vulcanizing agent Dicup 40	7	–	–	–	–	–	-"-
Zincum oxide	–	5	5	5	5	–	-"-
Phenil-β-naphtil amine	–	1	1	–	1	–	-"-
Resin Piccopal 100SF	–	13	6	–	3	–	-"-
Sulfur	–	1.5	1.5	1	1.5	–	-"-
Paraffinic wax	–	–	2	–	8	–	-"-
Vulcanization accelerators:							
Vulcafor DPG	–	0.25	–	–	–	–	-"-
Tetrone A	–	–	–	0.5	–	0.75	-"-
Tetramethyltiuramdisulfide	–	–	–	0.5	–	–	-"-
Di(2-benztiazol) disulfide	–	1.5	1.5	–	–	0.5	-"-
Vulcafor MBT	–	–	–	1	–	–	-"-
Vulcafor MBS	–	–	–	–	1.5	–	-"-
Triethanolamine	–	–	–	1	–	–	-"-
Diethylene-glycole	–	–	–	–	2	–	-"-
Aromatic naphtene oil Dutrex 95G	–	–	–	–	–	9.5	-"-
Heat shielding properties:							
Burning out time, sec	28.47	15.34	19.41	22.74	13.18	19.81	11.86
Erosion rate (average), mm/sec	0.23	0.43	0.33	0.28	0.53	0.29	0.52

According to the results of high temperature tests elastomeric HSM were found perspective also for internal heat shielding of MESP of multiple use, in particular the ones, based on polyurethane rubber, butyl rubber, etc. [8].

It is seen from the above considered, that various rubbers are used as the basis of HSM. But if consider the amount of publications, butadien-nitril, ethylene-propylene and silicon-organic rubbers are most of all applicable.

Physical properties of butadien-nitril rubbers, which are the product of copolymerization of butadien and nitril of acrylic acid, depend sufficiently on the content of the latter. The following actions occur in polymer with the increase of content of linked nitril of acrylic acid: intermolecular interaction between polymer chains increases sufficiently, as well as the density does, glass transition temperature increases, dielectric properties decrease, as well as solubility in aromatic solvents, and stability to swelling in aliphatic hydrocarbons increases also [9].

The processing of butadien-nitril rubbers is difficult because of high rigidity, stipulated by intensive intermolecular interaction. Processing ability of rubbers of different marks depends on initial viscosity and on nitril group content. Preliminary plastification is required for all rubbers of rigid types. In this case mechanical plastification of rollers at 30-40°C temperatures is mostly efficient.

Butadien-nitril rubbers may be vulcanized by sulfur in presence of sulfuric vulcanization accelerators. They also can be vulcanized by thyuram, organic peroxides, alkylphenolformaldehyde resins, chlorine organic compounds. Vulcanization is performed at temperatures of 140-190°C. Big vulcanization plato is observed during it. Vulcanization rate increases with the increase of the content of linked acrylic acid nitril. The most important cause of the wide application of butadien-nitril rubbers as the basis of HSM is the fact, that they can be vulcanized as well as to be well compatible with coking phenolformaldehyde resins. Coking products of such compositions possess comparatively high mechanical strength and erosion stability. However, elasticity of vulcanizates decreases abruptly with the introduction of resins.

Preliminary calculation of temperature on the border between the coating and shielding surface can be performed for heat-shielding coatings [10]. Thus, for the material, based on butadien-nitril rubber and filled by silicon dioxide, the calculation can be made according to the formula:

$$t_b - t_i = (t_d - t_i)^{(-Rx_i/a + R^2\theta/a)},$$

where t_b - final temperature of the coating surface, connected with shielding surface;

t_i - initial temperature of the coating;

t_d - decomposition temperature (for particular material it is about 150°C);

R - average ablation rate;

x_i - initial coating thickness;

a - temperature conductivity coefficient;

θ - heat influence duration.

Temperature, calculated according to this formula, differs approximately by 1.5°C from that, obtained experimentally. Making corresponding transformations, it may be determined necessary coating thickness for definite duration of engine exploitation and admitted increase of temperature of its wall.

Butadien-nitril caoutchouc based rubbers usually contain silicon dioxide or asbestos as the main filler. They are applied for the shielding of nozzle assembly elements, internal cylindric surfaces of combustion chambers of MESP, front and nozzle bottoms [11,12]. Thus, for shielding steel cover of MESP of "Titan 3-C" missile Thompson Fiber Glass Company has worked out the rubber of Hitca 6520, filled by silicon dioxide [13]. The cover is applied by adhesion of procurement from calendered rubber sheets with the following their vulcanization in autoclav at pressure of 21 kG/cm^2. Similar material, produced by Tyokol Chemical Corporation, is adhesed to the framework of the engine and vulcanized at 160°C during 17 hours and at pressure of 6.3 kG/cm^2, performed directly in the framework. It is closed hermetically by lids for this purpose [14]. Another variant for back lid of MESP of "Titan 3-C" missile is rubber, containing combination of silicon dioxide and asbestos as the filler [15].

Tests of highly filled rubber of similar composition, applied for heat shielding of the framework of the first section of "Minuteman" missile (cover thickness of 2.5 mm), showed, that the rate of its coking is 30% higher than that of the material, applied before [16,17]. It is pointed out, that the application of rubber allowed to decrease cover weight twice and to simplify its applying by adhesion of procurements to the framework and their vulcanization under pressure at the increased temperature. More perfect method is preliminary procurement vulcanization in press-forms, that provides high accuracy of sizes and form of cover sections and their full correspondence to the configuration of shielding framework parts. The vulcanization is performed under rather high pressures and thorough temperature regulation, because the rubber possesses low plasticity.

General Tire and Rubber Company has patented in Great Britain [18] and USA [19] heat shielding rubber of the following composition (in weight parts):

Butadien-nitril caoutchouc 100.0
Zincum oxide 5.0
Symmetric di-β-naphtilparaphenylenediamine 3.0
Benzothyazoldisulfide 1.0
Tetramethylthyuramdisulfide 0.05
Cumaron-indenic resin 12.5
Dioctylphtalate 12.5

Stearic acid 1.5
Asbestos fibers 40.0
Hydrated active silicon dioxide 20.0

Vulcanization of the composition was performed at 149°C during 60 min. During testing disks from this rubber, possessing oriented fibers, in oxygen-acethylene burner flame (distance from nozzle is 25.4 mm) temperature of the back side of the sample reached 204°C after 90 sec. Flame direction was changed from 60 to 90° according to sample surface with the frequency of 10 cycles per minute in order to imitate turbulence of the flow of propellant combustion products. In these conditions material remove was obtained 0.025 mm/min, i.e. it was sufficiently lower, that for the materials, based on butyl caoutchouc, plastified butadien-styrene rubber and methylphenylpolysiloxane elastomer, which remove reached 0.07 mm/min.

Another patent encloses rubber mixture, containing butadien-nitril caoutchouc (35-75%, better 40-50%), phenolformaldehyde resin and graphite (65-25%, better 60-50%) as fillers [20]. The rubber is prepared on rollers, for which it is mixed 100 weight parts of caoutchouc, 110 weight parts of resin and 15 weight parts of tricrezylephosphate. After that rollers are heated, and then 2 weight parts of tetramethylthyuramdisulfide are introduced after heating. Then rollers are cooled again, and 10 weight parts of hexamethylentetramine and 215 weight parts of graphite powder are introduced first of all. Components are carefully mixed. Intermediate product is molded into covers, in this case vulcanization regime is 135 min at 165°C. Strength limit at the material break is 312 kG/cm^2, relative elongation is 5%, hardness according to Shore (scale D) is 67; density is 1.45 g/cm^3, erosion rate is 0.017 mm/s. Erosion was determined in engine with propellant combustion temperature of 3315°C, initial rate of propellant combustion product of 275 m/s, pressure in combustion chamber of 21±3.5 kG/cm^2, engine exploitation duration of 31±6 sec. In the same conditions erosion rate of heat shielding asboplastic was 0.132-0.155 mm/s.

Combination of phenole-formaldehyde resin and boracic acid may be used also as the filler [21].

High erosion strength is shown also by heat shielding rubbers V-44 and ORSO-9250 in prenozzle zone of largescale MESP (22). These investigations were performed according to five and seven segment MESP of "Titan 3M" missile. Rubbers are filled by silicon dioxide and asbestos, respectively. The rate of cover erosion in dependence on heat flow for five segment MESP was found 2.5 times lower, than for seven segment one (from 0.25 to 0.5 mm/s in comparison with 0.5-1.25 mm/s at the change of flow by 2.5 times approximately).

Rock Island Arsenal Company worked out the rubber, combining phenole-formaldehyde resin and asbestos fiber fillers [23]. The density of this material is 1.34 g/cm^3, relative elongation at break is 40%.

The Laboratory of reactive motion of California Technological Institute of USA worked out the rubber of Gengard-52 mark, containing silicon oxide hydrate, asbestos and additives [24]. The material is used in MESP of

Earth satellite ATS. MESP parameters are: weight - 257 kg, length -1390 mm, diameter - 715 mm, nozzle expansion degree - 31: 1; propellant weight, based on polyurethane with ammonium perchlorate - 232 kg; pressure in combustion chamber - 18.2 kG/cm^2, engine exploitation duration - 45 s. Heat shielding cover thickness changes from 2 mm in cylindric part of MESP to 5 mm in nozzle range. Preparation of covers of variable thickness is performed by application of additional layers 0.75 mm thick on the initial 2 mm thick layer.

Paper [25] points out, that in small MESP HSM, based on butadien-nitril caoutchoucs, are applied on protecting surfaces as nonvulcanized intermediate product with its further vulcanization at corresponding temperature and pressure, whereas in large scale MESP of 6.6 m diameter, for example, it is accepted the method of the application of preliminary vulcanized sheets.

Rubbers, filled by asbestos and intended for heat shielding of MESP frameworks of 6.6 m diameter, was worked out by Goodyear Tire and Rubber Company [26, 27]. Rubber is delivered as intermediate products, representing large segments, treated by preliminary vulcanization is Verson presses; Intermediate product preparation is performed under careful interoperational control of technological process.

The cover is applied in special camera 45.75 m long and 18.3 m high with constant temperature (20°) and relative humidity (30%). Segments are placed along cylindric part of the framework, pressed by hydraulic winches to the walls and are exposed under pressure till the end of evacuation, adhesive hardening and complete rubber vulcanization.

The cover is applied to the front and nozzle lids, consisting with eight sectors, adhesed by epoxy adhesive and treated by the same processing, as framework cover. Segment junctions are filled by the composition of cold hardening of V-61 mark with the help of spatula. After cover application its borders are smoothed by polishing. High quality of the cover is reached owing to the application of single layer segments - intermediate products. Total weight of the cover is 4.5 tons (28-30).

Rubber, filled by asbestos, is applied for heat shielding of experimental MESP of 4 m diameter, produced by Lokhid Production Company (31, 32). Engine framework consists with the section, possessing front bottom (6.1 m long), central section (6.7 m long) and the section with nozzle bottom (5.6 m long). HSM is applied just to end parts of the framework and is of variable thickness, which reaches 44.5 mm thickness near butts. The cover of such thickness is intended for works at repeated engine startings; destroyed part of the cover is removed mechanically, in the case of single use its thickness is found 19 mm. Figure 1 shows the application of butadien-nitril rubber to internal MESP surfaces.

The investigation is considered to be important, which was performed in order to study migration of nitrogylycerin and dinitrotoluene from missile propellants NgR-005 and NgR-006F at the contact with rubbers, based on butadien-nitril caoutchouc with different concentration and type of vulcanizational bonds, in particular, with mono-, di- and polysulfide bonds

[33]. In this case it was cleared up, that high stability to the influence of missile propellants is shown by rubbers, based on butadien-nitril caoutchoucs, containing monosulfide cross bonds after vulcanization. Investigators think that intensive degradation of rubbers with caoutchoucs with di- and polysulfide vulcanizational bonds at the contact with propellants is caused by nitrogen oxides, formed during nitrocellulose decomposition.

Specific application of heat shielding rubbers, based on butadien-nitril caoutchoucs is in their usage in collapsing nozzles of MESP with high expansion degree [34]. MESP of Skybolt missile was used for model experiments. One of the variants of heat shielding cover is V-44 materials of about 9 mm thick.

The same material was considered as one of the competitors (others were heat shielding rubber JBT-10, silicozemphenolic composition FM 5131, graphite-phenolic composition MX 4926) for the application in nozzle block of MESP of NASA 260 SL-3 type [35].

Some patents, concurrent with the variants of heat shielding rubbers, based on various synthetic caoutchoucs, enclose on a level with them the materials, based on butadien-nitril caoutchoucs. Thus, the competition, based on polar elastomer with powder-like highly active filler silicon dioxide, or based on it and (or) with complex filler of mica type, vulcanized by organic peroxides or metal oxides, is suggested as the material for heat shielding coatings of MESP, possessing sufficient thermostability, good adhesion to shielding surfaces and good compatibility with solid-propellant grain (6). The following substances can be used as the basis of the material: 1) ethylene and vinylacetate copolymers, containing 35-60% of the latter; 2) butadien-nitril caoutchoucs, which may contain carboxile groups, for example, copolymers of butadien, acrylonitril and acrylic acid; 3) polychloroprene. Layered filler may consist 5-30% of elastomer weight, and together with powder filler - 120% of elastomer. To fix the sheets of preliminarily vulcanized rubber on shielding surface, adhesives are used. The example of composition recepture for heat shielding rubber (in %): 100 of copolymer of ethylene with vinylacetate, 40 of powder silicon dioxide hydrate, 4 of 40% dicumyl peroxide, 1 of stearic acid (dispergator), 0.5 of phenol-α-naphtylamine (antiger), 2 of trialkylcyanurate (vulcanization activator), 2 of polycarboimide (antioxidant).

One more patent [36] encloses the composition, based on hydrocarbonic or acrylonitril caouthchouc, containing dehydrogenation component - the composition of the element of 4-6 period of the VIII-th group. The composition must not contain sulfur or its compounds as vulcanizing agents in order to exclude the deterioration of its properties. Silicon dioxide, asbestos or carbon are the fillers of the composition. The application of the catalyst allows to increase sufficiently the outlet of coke residue (≥10%). Isobutylene copolymers may be used as the basis of the composition, containing ≤5 molar % of butadien, ethylene with propylene, styrene with butadien, butadien with acrylonitril. For example, 100 weight parts of isobutylene copolymer with 0.9 molar % of isoprene are mixed with 25 weight of the component, containing 67% of nickel, and with 25 weight parts of silicon dioxide.

Obtained material is treated by pyrolysis in high-frequent furnace with graphite element, allowing to heat the sample up to 1100°C during 1 min. Outlet of coke residue and gas hydrogen for the composition, enclosed by the patent is 45% and 8.2%, respectively, whereas for the composition, based on the same copolymer, but containing only 50 weight parts of silicon dioxide, these indexes equal 1% and 0.9% respectively.

Sufficient interest is presented by HSM, based on mixture of butadien-nitril and ethylene-propylene caoutchouc [37]. Marks of the materials, based on mixture of these caoutchoucs, are 370-9966 and 370-9966A. Such materials possess high heat shielding properties. Thus, at testing of samples 3.5 mm thick, heating by quartz lamp, after 60 sec temperature of the surface was 428°C, whereas back side of it was heated up to 115°C, only.

Ethylene-propylene caoutchouc is of great interest for the application in HSM in caoutchouc mixtures, as well as their basis, that is explained by properties of this caoutchouc. If the caoutchouc chain contains over 15% of propylene, copolymers of ethylene and propylene display rather high elastic properties as a consequence of crystallization difficulty.

If 30-50 molar % of propylene is contained in the chain, then the polymer becomes completely amorphous and possesses the highest elasticity [9, pp. 87-91].

To obtain polymers, able to be vulcanized by sulfur, ethylene and propylene are copolymerized with dienic monomer, obtaining triple ethylene-propylene caoutchoucs. The content of the third monomer in caoutchouc is 0.3-2 molar %. The increase of content of unsaturation in caoutchouc causes the increase of vulcanization rate and vulcanizate strength.

Physical properties of ethylene-propylene caoutchoucs depend on the ratio of ethylene and propylene in the polymer and do not depend, in fact, on the existence of the third monomer. Caoutchoucs, delivered by the industry, contain 30-40% of propylene and do not crystallize neither during storage nor at deformation.

The density of caoutchoucs is 0.85-0.87 g/cm^3, glass transition temperature is 58-65°C. Caoutchoucs are stable at mechanical and thermooxidative plastification. Rubber mixtures possess good molding ability. Molded intermediate product articles retain well their form because of high thermoplasticity of the caoutchouc. Rubbers, based on ethylene-propylene caoutchoucs, are widely applied for the preparation of the articles, exploited in aggressive mediate at temperatures up to 150°C. They are used in the production of rubbered fabriques, flexible pipes, thermostable belts, molded and nonmolded machine parts, cable isolation, etc.

Rubbers, based on ethylene-propylene caoutchoucs, are preferable as HSM in the cases, when fiber glass frameworks of MESP are prepared by winding, in particular, for Cevlar material with epoxy linkage, because of high elasticity and good compatibility with the framework [38, 39].

Combustion chamber diameter is 762 mm, engine length is 635.4 mm. Total volume is 212 l. Heat shielding material, based on triple ethylen-propylene caoutchouc, is of GSM 55 mark, its density is 1.1 g/cm^3. Silicon dioxide was used as the filler. Applied coating thickness is from 3.56 mm on cylindric part of shell up to 14.7 mm on the nozzle part. Total weight of the heat shielding is 15.95 kg. HSM possesses high erosion strength, because applied propellant UTP-19.697 is mixed one and contains 20% of aluminum. Similar material was applied in SRM-1 MESP construction [40].

It was tested 14 modifications of HSM, based on ethylene-propylene caoutchouc, in the flame of oxygen-acethylene burner in model MESP in solid-propellant combustion products. The compatibility with solid-propellant grain was investigated also [42]. Asbestos fiber was used as the filler of the caoutchouc. Optimum recepture R-4 is used in MESP, working according to the following parameters: combustion duration - 3.7-3.8 sec, average pressure in combustion chamber 3.7 MPa, linear coating remove - 0.88-1.23 mm.

There are reports about the application of ethylene-propylene caoutchouc for HSM for cosmic articles, also [42, 43]. Thus, Italian "Shia Viscoas" Company worked out MESP for European artificial Earth Satellite GEOS, providing its transmission from intermediate to stationary orbit. The engineer must work in temperature range from -10 to 40°C at axial overloadings of 16g, acceleration vibrations up to 10.2g, rotation with frequency of 123 min^{-1}. Engine diameter is 684 mm, length is 1130 mm, its mass is 303 kg, propellant mass is 267 kg, specific pulse is 286 kG·s/kg, maximum flight thrust is 2360 kGs, average pressure in combustion chamber is 23 kGs/cm^2, exploitation duration is 47-51 s. HSM is filled also by asbestos and provides temperature of shielding surface less than 250°C after finishing MESP work. Heat shielding weight is 7.9 kg. Rigidly fixed cartridge consists with 14% of polybutadien with circle carboxile groups, 76% of ammonium perchlorate and 10% of aluminum. In this connection HSM must possess erosion stability.

Another MESP is worked out by Tyokol Chemical Corporation for space vessels, intended for investigation of planets, dislocating behind Mars, for example, space vessel, launched to the orbit among Jupiter [44]. Light material, based on phenolic resin, is supposed to be used as heat shielding. It will be filled by cork. To prevent cork chipping coating is covered by rubber, based on ethylene-propylene caoutchouc of Tire and Rubber Company. This very combination of two materials was found the most efficient heat shielding during tests.

There exists a list of HSM, based on ethylene-propylene caoutchouc, which are enclosed in patents.

Thus [45] shows the enclosing composition, consisted with 80-100 weight parts of copolymer of ethylene, propylene and dien with nonconjugated double bonds, 5-20 (preferable 10) weight parts of polybutadien with end COOH-groups, 40 weight parts of perchlorinepentacyclodeacane and 20 weight parts of antimony oxide to increase combustion strength, 75-150 (preferable 100) weight parts of asbestos and 7.5 weight parts of 2.5-dimethyl-2.5- (tret-butylperoxi) hexane as vulcanizing agent. For example, 100 weight parts of copolymer of ethylene with propylene is added by 100 weight parts of asbestos and 10 weight parts of polybutadien with end COOH-groups. They are mixed on rollers during 45 min, approximately. Obtained mixture is added by 7.5 weight parts of 2.5-dimethyl-2.5-di (tret-butylperoxi) hexane. This combination is mixed during 5 min, approximately. Then the mixture is added by 20 weight parts of antimony oxide and 40 weight parts of perchlorine-pentacyclodecane, and rolling is continued during 10 minutes, approximately, until homogeneous composition is obtained. Temperature of mixing is 107°C. Obtained plain intermediate product is cut into sheets of the required size and is vulcanized at 171°C during 30 minutes under pressure of approximately 31 kGs/cm^2. The density of hardened HSM $\rho \sim 1.4\ g/cm^3$, hardness according to Barckol is 32-42. Strength limit at break of HSM is 9.8 kGs/cm^2, relative elongation is 36%, modulus is 57.89 kGs/cm^2 and hardness according to Shore is 77 after additional second stage of vulcanization at 121°C during 10 hours, approximately. To determine coking the samples of 50.8*50.8*5.08 mm size, treated by the second vulcanization, are displaced in MESP, working on propellant with combustion temperature of 3204°C at the rates of solid propellant combustion product flow of 20.09 and 137.9 $g/cm^2 \cdot s$. In this case the rate of coking of HSM samples at 20.09 $g/cm^2 \cdot s$ was found 0.2794 mm/s, and at 137.9 $g/cm^2 s$ - 0.762 mm/s.

Other patents [46, 47] enclose the materials, based on copolymer of isobutadien and isoprene, butadien-styrene, chloroprene and other caoutchoucs, concurrent with HSM, based on ethylene-propylene caoutchouc.

HSM, based on chlorine containing caoutchoucs, has also found application in MESP [48].

French patent encloses HSM, based on synthetic chlorinated caoutchouc (chloroprene preferably), containing 25-100 weight parts of thermostable fiber (silicon dioxide, asbestos, etc.) and various additives per 100 weight parts of caoutchouc. For example (in weight parts): neoprene WRT - 50, asbestos - 50, antiager Permanax 45 - 1, stearic acid - 1, magnesium - 4, zincum oxide - 5, ethylene-tiocarbamide - 0.5. After mixing on rollers rubber mixture is calendered to the thickness of 2 mm, required sheets are cut from intermediate product according to the form of protecting surface. After that molding of coatings is performed during 30 minutes at 160°C.

Tyokol Corporation has worked out heat shielding coatings, based on various caoutchoucs. Thus, the material, based on polyisoprene with asbestos and carbon black as fillers, possesses sufficiently high mechanical strength at elongation (1120 kGs/cm^2) and heat shielding characteristics: at testing in the flame of oxygen-acethylene burner (distance between sample and nozzle is 19 mm, inclination angle of the flame is 45°) mass remove was found 20% after 60 sec., ablation rate - 0.135 mm/s, and temperature of back side of the sample increased by 12°C, only [49].

Nonvulcanized material is produced as calendered sheets 1.5 m wide and 0.25, 0.75, 2.0 and 2.54 mm thick or as cyldrindic procurements.

Rubbers, based on butyl- and polyurethane caoutchoucs, were developed for internal heat shielding of MESP of multiple starting. The coatings were tested successfully at multiple (up to 12 including) starts of the engine [8].

Rubbers, based on polyurethene caoutchoucs, may be applied to heat shielding coatings, working in the flow of solid propellant combustion products, possessing combustion temperature up to 3300°C, forming stable coking layer under the influence of high temperatures. The expense of such coatings is 22 dollars per 1 kg [50].

USA patents contain the information about the fillers, unusual for rubber industry, applied in HSM. Thus, patent [51] and notes [52, 53] enclose the material, containing chromium oxide as the filler. Patent [54] encloses metal haloid (zincum chloride), possessing boiling temperature of 500°C and high pressure of vapors in the range of 500-1200°C.

There have been worked out the materials, also, possessing increased erosion stability with freezing stability to -40°C [55]. They are produced on the basis of elastomer, participating in the composition of solid propellant. That is why they are interesting for the application in the cases, when propellant contacts directly with heat shielding in the engine, i.e. the gain is strongly fixed. The application of such materials is mostly advisable in engines of big sizes. In this case the coating is applied to the walls as partly vulcanized procurements, which complete vulcanization is performed simultaneously with elastomer, on which the propellant is based.

Materials, produced by Goodreach Company, are intended for exploitation at temperature up to 3850°C and propellant combustion duration up to 2 min [56, 57]. It is pointed out the expediency of the application of coatings, expensed during propellant combustion.

It is also informed about the fact, that rubber-like HSM may be applied in wide range of temperature (1360-2760°C) and pressures (35-140 kGs/cm^2) [58].

Figure 2 shows multilayer rubber coating, applied to spheric lid of large MESP [59]. Specific feature of this coating is smoothly changing thickness, that provides true shielding of various parts of the framework according to the intensity of the influence of the flow of solid propellant combustion products. Rubber is applied by layers, which number reaches 20 on parts, treated by maximum erosion. It is sufficiently important to exclude noncementation and to remove the solvent from the adhesive. Vulcanization is performed in autoclav, filled by inert gas at the increased temperature and moderate pressure.

There is not a lot of information rubber-fabrique (reinforced) elastic materials, based on organic caoutchoucs, despite the fact, that they possess the increased erosion strength in comparison with rubbers. It is pointed out [60], that coking rubber-resin compositions of United States Rubber, reinforced by fabriques, show higher erosion strength, that plastics, at comparatively lower density; temperature range of their working ability is wider, than for plastics.

Apparently, "Astracan" material, based on Hitco-158 mark rubber and Pliofen 5900 resin, reinforced by carbonic fiber and worked out specially for nozzle systems of some engines [61], can be related to similar systems. The material possesses the density of 1.4±0.05 g/cm and is vulcanized during 4 hours at 160°C and 150 bar pressure.

In some cases special demands are made to HSM. Thus, for example, sometimes HSM must protect blocks from the appearance of splits in them during mounting in engine framework as well as to provide safety of MESP framework from high temperature flows of solid propellant combustion products. In this case it is important high elasticity of HSM for the exclusion of breakdowns in work or MESP failure. In this case it is suggested heat shielding coating, consisted with two thin layers of rubber, based on natural or synthetic caoutchouc, between which perforated sheet with holes filled by the same rubber is dislocated [62]. Such coating may be deformatted at shrinkage of blocks and is nonpressed under pressure, formed during propellant combustion. The above mentioned two layers of rubber are adhesed to each other not by the whole contact surface, and by separate parts only, possessing the form of longitudinal belts, which width is approximately 3 times lower, than that of intermediate longitudinal belts, by which adhesion is not performed. Such connection in zones, which are not adhesed, is provided by laying silicon and polytetraethylene belts. Rubber layers are preserved from adhesion by lac application, also. Such heat shielding system preserve solid-propellant block from the influence of vibration, centrifugal forces, etc. from the side of MESP framework. Emptinesses, connected with each other, are formed between adhesed zones at coating deformations, that allows to distribute pressure uniformly along all engine circuit.

Wide application for heat shielding of internal surfaces of MESP and direct-flow missile engines (DME) was found by HSM, based on silicon-organic elastomers, because these materials work successfully under the influence of restoring medium, that takes place in MESP and on the 1st start stage of DME work, as well as in oxidative one, that is performed in DME on the 2nd marching stage of the work. This is explained by features of thermal decomposition of silicon-organic elastomers.

Comparing with hydrocarbonic caoutchoucs, possessing equal energies of carbon-carbon bond (58.6 kcal/mole) in the main chain, silicon-organic elastomers possess sufficiently higher bond energy of silicon-carbon type (from 101 to 118 kcal/mole) in the main chain, calculated from combustion heats [63]. Polar silicon-carbon bond stipulates high lability and ability for regrouping at thermal influence. That is why sufficiently important property of silicon-organic elastomers for the application as the basis of HSM is their ability to form cyclic low molecular compounds, possessing high thermoabsorption, under the influence of high temperatures [64-66]. On the whole, cyclic trimer is formed during the decomposition of dimethylpolysiloxane elastomer [67].

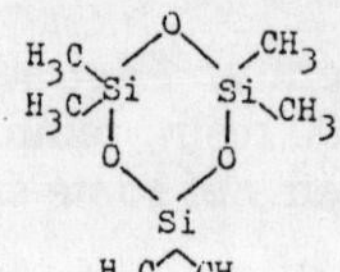

Besides, this timer, which content reaches 44%, the following dimethylcyclosiloxanes can be formed at polydimethylsiloxane heating over 400°C in vacuum: $[(H_3C)_2SiO]_4$ - 24%; $[(H_3C)_2SiO]_5$ - 9%; $[(H_3C)_2SiO]_6$ - 10%, $[(H_3C)_2SiO]_{n>6}$ - 13 % [68]. Further decomposition of cyclosiloxanes into low molecular products requires also sufficient amounts of heat energy.

The ability to high temperature transformations of silicon-organic elastomers into cyclic compositions is stipulated also by molecule structure [69]. Molecules of polydimethylsiloxanes possess spiral-like form with six or eight chain units in single loop of the spiral as a consequence of high flexibility of molecular chains and low sizes of methyl groups. The influence of high temperatures enforces heat oscillation of separate parts or units of molecule chain. Polysiloxane chain break occurs by silicon-oxygen bond. In this case spiral-like structure of the molecule creates successful conditions for closing parts of the molecule into cycles in the places of their ends coming close after the break.

The experience of work with organic polymers shows concretely, that one of the most important conditions of their coking is the formation of cyclic structures in changing material. In opposite to organic cycles, among which 5- and 6-unit ones are most stable, under the influence of high temperatures on silicon-organic polymers in inert medium it is possible the formation of cyclic polymers with low molecular weight, and in the presence of oxygen - polymeric silicon oxides, which typical representative is silicon dioxide $(SiO_2)_n$. For example:

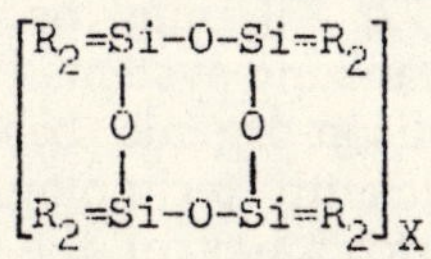

In this case each silicon atom is linked with four atoms of oxygen, whereas carbon does not form similar stable structures. This is explained by the fact that

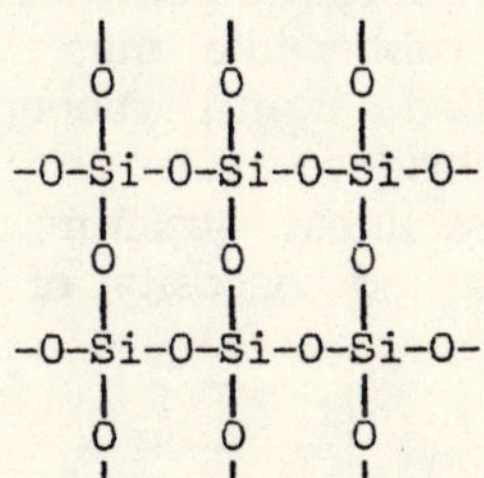

silicon, possessing larger atomic size and higher charge than carbon, forms stable polymeric formations, and carbon - thermally instable ones. Thus, for example, specially synthesized polymers - polyoximethylenes with similar molecule structure $-CH_2O-CH_2O-$ are thermally unstable and can decompose easily already at 170°C.

Higher energy of silicon-carbon bond in comparison with the energy of carbon-carbon bond allows to suppose, that evidently, higher energy flow expenses are required for decomposition of polymeric compositions of $[R_2SiO]_m$ and $[SiO_2]n$ type in comparison with carbonic cokes. Moreover, at high temperatures materials, based on polysiloxanes, form also carbonic residue, which outlet depends sufficiently on chemical structure as well as on applied fillers [70].

The important feature of the compositions, based on silicon-organic elastomers is the absence of aging in sufficiently strong climatic conditions, that explains their high strength to the influence of oxygen and nitrogen. This is stipulated by the absence of double bonds in chain molecule of the elastomer [71]. Precisely the absence of double bonds leads to the fact, that elastomer does not enter addition reactions. After vulcanization silicon-organic compositions are completely nontoxic, possess good stability to the influence of oxidating agents, solutions of salts, acids, alkali, and high dielectric properties, also.

Vulcanizates of dimethylpolysiloxane elastomer, despite carbonic ones, possess elastic properties in sufficiently wide range of temperatures (from -70° to 300°C), that is connected with large interatomic distance in silicon-oxygen unit in comparison with carbon-carbon bond [9, p. 89].

It should be mentioned, that vulcanization of compositions, based on silicon-organic elastomers, according to methods, worked out for the first time by N.B. Baranovsky et al. [72, 73], may be performed without heating. Further on other classes of vulcanizing systems [74-79] were developed beside tin containing ones. Thus, silicon-organic based HSM, being vulcanized without heating, must be sufficiently technologically efficient than the ones, based on organic rubbers. Shortcomings of the latter are the necessity of heating up to 150°C and solvent remove - in some cases.

Silicon-organic linkages represent great interest for rubber-fabrique materials as well as for the ones, reinforced by cells also, because elastomers are produced not only as rubber-like material, but as liquid flowing consistence, which are called liquid rubbers. The latter sufficiently simplifies impregnation of fabrique carcass and cellpast filling [79]. Liquid silicon-organic rubbers possess linear structure, molecular weight - in the range from 20,000 to 100,000 and viscosity of 0.5 - 80 Pa. Their general formula is:

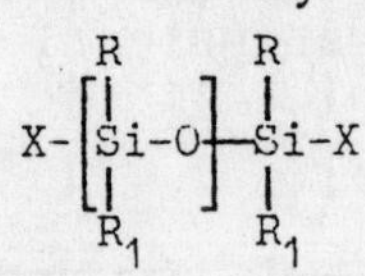

where R and R_1 - methylic, phenylic, vinylic and other radicals; X - functional groups, for example OH [80].

Methods of obtaining linear polysiloxanes are described in [81-83].

The attempts of developers of silicon-organic elastomers are aimed, generally, to the increase of freezing, heat-, thermo- and oil stability [84]. It was found, that the increase of freezing stability may be provided by suppression of the ability of polyorganosiloxanes to crystallize. For this purpose regularity of their structure is broken at the introduction into side surrounding of methylphenilic, diphenilic and diethylic groups, beside dimethylic ones [85-88]. Most perspective way of the creation of freezing stable siloxane elastomers is the obtaining of polydimethyldiethylsiloxanes, possessing glass transition temperature in the range of -135 ÷ -140°C.

Heat- and thermostability of silicon-organic elastomers is increased by the introduction of aromatic side groups into their structure [89, 90]. Another way of the increase of these properties is the synthesis of polyblock copolymers, consisting with rigid and elastic blocks [91].

Polyorganosiloxanes are also characterized by hemo-, humid-, ozone, atmosphere and vacuum strength, stability to the influence of radiation (especially elastomers with phenilic groups), fungi resistance, good coking ability [92-94]. They differ by high capacity for various fillers (up to 100 weight parts of filler per 100 weight parts of elastomer), form foams easily and may be reinforced by fabriques and cells.

Silicon-organic materials for MESP show sufficient advantages before HSM, based on polychloroprene, butyl caoutchouc, polyurethane [95], fluorine caoutchoucs, polysulfides, epoxy resins [96], and before phenol-neilon erosion stable material, also [97-101].

The conclusion about the advantage of silicon-organic materials with organic and mineral fillers was made after the performance in tests in high temperature gas flow of solid propellant combustion products, flowing from MESP at high rates. The investigations showed, that erosion resistance of the materials is sufficiently influenced by chemical nature and amount of the filler, which can be changed from 10 to 100% in dependence on coating application. The higher intensity of heat- and mass transfer and erosion influence of the flow is, the large amount of the filler is required. Size and geometry of coatings is defined by the behavior in the flow and construction ideas.

Silicon-organic rubbers have found wide application for shielding combustion chambers and gas passes from combustion products of solid propellant in missiles of "Minuteman", "Bomark", "Polaris", "Saturn", "Titan IIIC" type, etc. [102-106]. In this case the weight of silicon-organic heat shielding, for example on "Minuteman" missile, is 250 kg, and that of the first jog of "Polaris" missile - 50 kg [107].

One of such silicon-organic rubbers is USR material, filled by phenolic resin and inorganic hydrates [108]. Its properties are shown below:

Density, g/cm^3 1.24
Strength limit at break (at 21°C), kGs/cm^2 56-84
Relative elongation at stretching (at 21°C), % 200-400
Hardness according to Shore (scale D, at 21°C) 45
Heat conductivity coefficient (at 93°C), W/m*grad 0.26
Specific heat capacity (at 65°C), kJ/kg*grad 2.3

A little higher density is possessed by silicon-organic rubber AMS 3345-1.32 g/cm^3. This rubber was applied in "Minuteman III" missile [109]. Heat shielding coating thickness is 1.65±0.13 mm. High temperature tests of samples of 38 mm diameter were performed on stand reactive engine at pressure in combustion chamber of 7 -12.2 kGs/cm^2 and dislocated at 48.3 cm distance from engine nozzle. The tests were performed according to stage regime: 4.2 sec at heat flow of 4800-5480 kJ/m^2, 1.3 sec at 798-1370 kJ/m^2. Table 3 shows the data on the material moving away.

Table 3. Erosion strength of AMS 3345 material

Exposition duration, s	Heat flow, kJ/m^2	Material moving away	
		g	mm
4.2	4800-5480	1.00	1.15
3.0	2970-3310	0.97-1.17	0.75-0.90
1.3	798-1370	0.19-0.43	1.52-0.33

The material is applied to the shielding surface through the support from fiber glass fabric. It performs heat shielding of internal surfaces of PBPS engine from the influence of high temperature flow with solid particles during engine performance and separation of the third section of the missile.

Another preferable material for nozzle block is the composition on the basis of two-component hermetic, based on dimethylpolysiloxane elastomer DC 93-022 of Dow Corning Company, which is hardened at the increased temperature [110]. Further investigations of the material modifications were performed at the value of heat flow of 8631, 11858 and 17598 kW/m^2 at the rate of high temperature gas flow up to 3.0 M. The best results were obtained at the substitution of dimethylpolysiloxane by methylphenilpolysiloxane, filled by silica carbide (the following substances were tested: aluminum oxide, boron carbide, boron nitride, molibdenum carbide, niobium carbide, titanium carbide, zirconium boride, tungsten carbide, boron nitride, zirconium carbide, zirconium silicate) and reinforced by carbonic fibers. The material was marked as 93-104, its heat conductivity coefficient is 0.33 W/m*grad. The material 93-104 may be reinforced by asbestos or polyimide, aromatic polyimide, polyester fabrics, also in order to broaden the range of its application in more strong conditions of work. Reinforcement increases shielding properties of polymeric material [111]. Antioxidants and antiozonants may be additionally introduced into the material. The same material may be used for the preparation of coatings by rolling methods. For this purpose it is performed the reinforcement by steel network belts. Necessary orientation and high quality of the surface is provided as a result [112].

Synthetic fibers are interesting from the point of view of reinforcing filler, wood fillers were interesting as powder ones [113].

For shielding cylindric internal surfaces of MESP, working on strongly fixed slotted grains, DC-93-118 material of Down Corning Company is intended. But for these conditions other materials may be applied also, produced by this company, made on the basis of silicon-organic elastomers [114].

Enterprises-customers receive silicon-organic rubbers as calendered sheets, belts or extrused fabrics [115]. In this case enterprise-deliverer is informed by customer about thickness, width and delivery conditions (vulcanized or nonvulcanized products). Rubbers are of complicated recepture, beside elastomers and vulcanizing systems they contain powder-like fillers, fibers, fabrics or filaments in order to provide the relation to demands to coatings for the articles. As a rule, such materials are applied in MESP with low rate high temperature gas flows of products of solid propellant combustion.

Vulcanized mixtures are stored in closed containers at 25°C without any change during 6 months at least. Molded coatings possess longer storage time. Pressure is not required for vulcanization, but small pressure is necessary for the application of fluidity, material densing and coating molding. Vulcanization is performed at 160-190°C. Vulcanization duration depends on temperature and configuration of rubber articles. The plate for physicomechanical tests of samples is vulcanized during 15 min at 160°C at its thickness of 1.6 mm. The second stage of vulcanization at 175-200° allows to remove side reaction products. No unusual or extraordinary safety measures are required at the processing of these mixtures.

Silicon-organic HSM are produced also as molding compositions, filled by short, irregularly dislocated fibers. Such compositions can be processed into articles of complicated form without heating under pressure. Articles, possessing high mechanical strength in low rate high temperature gas flows of the products of solid propellant combustion may be obtained by the selection of molding conditions at pressform filling.

Sufficiently strong coke-like layer is formed on the surface of the material during its decomposition. This layer improves working parameters of coatings.

Compositions, based on silicon-organic elastomers, can be applied as linkage in rubber-fabric materials. Carbonic, silica, quartz, aramide and other fabrics are used as fabric carcasses. For this purpose the same processes are suitable, which are used for the production of the materials from fabric, impregnated by organic resins. However, demands to vulcanization cycles and pressure (in autoclavs) may be simplified and decreased. Materials possess high technical characteristics, if produced from carbonic fabrics, as a consequence of stable coke formation. Materials with silica and quartz fabrics perform better work in oxidating atmosphere, for example, in the flow of fluid reactive engines.

The main material among silicon-organic elastomers is the chippest polymer polydimethylsiloxane. It can be modified by vinylic, phenilic and (or) fluorine-containing groups. These modified elastomers may possess increased oil stability, decreases crystallization temperature, high mechanical strength and low after constriction. The following materials are usually applied as vulcanizing agents of rubbers: organic peroxides of bis-2,4-

dichlorbenzoile, dibenzoile, dicumyle and di-tret-butyl - in dependence on polymer type and technology of its processing. The following materials are used as fillers: powder glass, silica, quartz, carbon, silicon carbide, ferrum oxide. Beside these commonly used fillers there exists the list of other materials, providing specific demands.

Development of receptures of silicon-organic rubbers is performed with the help of cylindric samples 30 cm long and initial diameter of 1.5 cm. They are tested in standard solid-propellant device 81.4 cm long and 15.2 cm diameter.

Mixed propellants, containing 20% of aluminum, are used for the tests. Average pressure in combustion chamber is 53 atm., test duration is 8.8 s. The results of tests of HSM, based on various materials, are shown in Table 4.

Table 4. Erosion stability of HSM

Material basis	Filler	Density, g/cm^3	Ablation rate, mm/s
Polydimethylsiloxane elastomer	Carbon	1.16	0.020
Ethylene-propylene rubber	Aramide fiber	1.14	0.025
	Asbestos fiber	1.28	0.015
	Asbestos fiber	1.23	0.030
Butadien-nitrile rubber	Carbon	1.24	0.015
Propylene rubber	Carbonic fabric	1.23	0.007
	Carbonic tricot	1.33	0.006
	Silica	1.45	0.019
	Quartz	1.49	0.020
	Carbonic fabric	1.48	0.005
Phenolic resin	Silica	1.60	0.020

Thin metal foil is adhered to the surface of silicon-organic rubber in order to prevent diffusion of polybutadien propellant components into it, for example, vulcanizing isocyanate agent [116].

The usage of foil allowed to solve the problem of application of silicon-organic rubbers in solid-propellant jet engines, performing on strongly fixed grains, because provision of adhesion of silicon-organic materials to the ones, based on organic rubbers, is sufficiently difficult. There is no difficulty in adhesion of both materials to the foil. But the introduction of intermediate foil layer is additional labour-consuming technological operation. For the first time the problem of performance of direct adhesion of such heterogeneous materials was described in patent [117]. For this purpose silicon-organic elastomer is combined with polybutadien, possessing hydroxile and groups, migrating to the coating surface. Propellant is also prepared on the basis of polybutadien with hydroxile end groups and with isoforondiisocyanate. Long-chain polymer is formed as a result of their interaction, which is insoluble in silicon-organic material and at the same time links it with propellant by isocyanate. In this case hydroxile groups of butadien do not interact with functional groups of silicon-organic elastomer. Technologically it is performed as follows. Polydimethylsiloxane, possessing hydroxile end groups, which average molecular weight is 2100, and percentage content of hydroxile groups is from 1.0 to 1.5, was heated at 70°C in vacuum with small amount of dibutyl dilaurate.

Two parts of obtained product were heated at 75°C during 24 hours with one part of polybutadien with hydroxile end groups. After that propellant was prepared, based on polybutadien rubber with hydroxile end groups, hardened by isoforondiisocyanate. The ratio of isocyanate groups and hydroxile ones is 0.8. Propellant was applied to the coating, based on silicon-organic elastomer, and then hardened. Physico-mechanical tests showed, that mechanical strength limit at coating detachment from the propellant was found 3.1 kGs/cm^2, whereas in the case of the coating with no additive this index is 0.73 kGs/cm^2, only. According to the opinion of patent's authors, the connection between coating and propellant is of chemico-mechanical character.

Insulrite material, based on dimethylpolysiloxane elastomer, was worked out for the application in nozzle block of JCJ Company missile engines. It was filled by silicon dioxide. There may be performed additional introduction of carbonic and aramide fibers, and reinforcement by fabrics (including tricot fencing) from these fibers and the ones of silicon carbide and quartz can be performed also. Molding of single-layered preparation is performed at temperature of 163°C during 6 hours at gradual pressure increase. Density of the material is 15% lower, than that of phenolic-silicon-organic compositions.

It was found in high temperature tests, that linear moving away of the coating is 1 mm/s and is regular without coating separating into layers, on which surface coke layer is formed. Under this layer the material preserves its plasticity [118].

Sufficient technological advantage of silicon-organic compositions, based on liquid elastomers, is the opportunity of the formation of the coating by centrifugal method or by pouring into molded cavity under pressure [119]. Such methods are applied for the preparation of coatings from the materials of Dow Corning Company DC-93-104, DC Sygard 184, DC E691-22E and General Electric Company - RTV-560 for internal surfaces of MESP and DFME, performing at alternate influence of high temperature flows at starting in restoring medium from products of solid propellant combustion and on march - in oxidating medium, when the air feeds combustion chamber.

Comparative high temperature tests of HSM, based on silicon-organic elastomers and phenolic asbo- and fiber glass plastics [120-122] were performed during finishing off missiles of "air-air" class of wide range (ALRAAM project) and strategic missiles of Navy and Air Force (the project of supersonic alate missile ASALM for firing to air and ground aims). In this case it was found, that the best heat shielding at the influence of the flow with 1980°C temperature are shown by DC 93-104 material. Temperature field distribution in this material correlates well with calculated characteristics of the materials.

The solution to apply silicon-organic rubber for missiles of "ground-ground" and "ground - air" class with missile direct flow engine Seram was decided by Martin Mrietta, also [123].

For long duration performing (up to 9 min) direct flow missile engines under 2080 K temperature the opportunity was considered to use rubber-fabric material with silicon-organic linkage, fabric carcass being made from three-dimension volume-strengthened fabric [124].

The application of two types of materials for shielding combustion chamber of direct flow missile engines is pointed out in the paper [125], also. One type is rubber-fabric materials with silica carcasses, treated by phenolic, epoxy, polyimide and other resins and impregnated by silicon-organic linkages. However, for these materials there are characteristic tendencies to the increase of heat conductivity at linkage coking under the influence of high temperature flow. In this case it is preferable rubber-like materials (again, for example, DC 93-104), because they are stable to the influence of high temperature oxidating flow, at the same time heat conductivity of coke layer being sufficiently lower.

To provide safe heat shielding of afterburner is ASALM missile, silicon-organic rubber is reinforced by cells from stainless steel [121]. The material possesses working ability under the influence of the flow with 1980°C temperature.

Materials, based on silicon-organic elastomers, are selected also for heat shielding coatings of engines for invisible tactical missiles [126]. Important feature of the basis of the materials is the fact, that plastifiers are not absorbed from the propellant (especially, nitrogen-ester ones), that decreases sufficiently smoke formation. The dislocation of launcher may be disclosed by the smoke, coming out from missile MESP. Moreover, if smoke is observed from far distance, than it is probable to save the target, to which the missile is launched. At visual guiding the missile, smoke decreases it visuality, that make it difficult to control its flight. The problem of the decrease of smoke formation can be solved in two ways. Beside silicon-organic based HSM double-based propellants, on which MESP are performing, are added by inhibitors, also based on silicon-organic elastomers. To obtain good connection between such additives and the propellant two products are used. The first one consists with elastomer, powder fillers (silicon and ferrum oxides) and vulcanizing agent. The second consists with the catalyst, containing tin salt, solved in liquid silicon oil. The ratio of products is 10:1.

At the development of HSM, based on silicon-organic elastomers, beside model stand MESP, there are methods of determination of thermomechanical properties. For this purpose there have been worked out the devices for pyrolysis with further identification of volatile products and solid residue with the help of mass-spectrography and γ-diffraction technique [127]. With the help of these devices there are detected the curves of temperature change according to sample depth in dependence eon test duration, heat conductivity and specific heat capacity of coke and initial material, heat of formation and the order of kinetic reactions.

CONCLUSION

1. Elastomeric HSM found sufficiently wide application for shielding surfaces of small- and large-scale MESP, working on both installed and strongly fixed grains, for missiles "Minuteman," "Polaris," "Titan," "Saturn," etc.

2. Heat shielding coatings are produced from intermediate products of monolyth and foamy rubbers or syntactic materials (with empty microspheres). They may be reinforced by fibers, fabrics or cells.

3. Elastomeric HSM possess high exploitation characteristics and do not split at deformations of shielding surfaces, heat and mechanical blows and vibration, during storage, transportation and engine performance.

Figure 1. The application of heat shielding rubber, based on butadien-nitril caoutchouc, to internal cylindric surfaces of MESP.

Figure 2. Multilayer rubber coating, applied to the lid of large-scale MESP.

References

1. *Missiles and Rockets*, v. 6, No. 19, p. 21, 1960.

2. *Rubber World*, v. 153, No. 2, p. 89, 1965.

3. S.B. Eglin, A.L. Landis, *Eight Quarterly Report Roc./Lin-BQ Hugles Aircraft Co.*, Culver City, California, 1962.

4. *Aerospace Management*, v. 41, No. 11, p. 21, 1961.

5. *Journal of Spacecraft and Rockets*, v. 3, No. 9, p. 1434, 1966.

6. *France patent* No. 2098934.

7. *Journal of the IRI*, v. 3, No. 1, p. 26-29, 1969.

8. W. Bradly. Internal insulation for multiple restart motors. *AIAA Paper*, No. 500, p. 8, 1968.

9. F.F. Koshelev, A.E. Kornev, A.M. Bukanov. *General Technology of Rubbers*, Moscow, Khimia, p. 72-76, 1978.

10. *AIAA Journal*, v. 2, No. 8, p. 1475, 1964.

11. *Aviation Week and Space Technology*, v. 81, No. 3, p. 66, 1964.

12. J.H. Daly, W.A. Harts, D.A. Meyer, J.G. Sommer. Nitrile-butadiene rubber in ablative appliation. *Appl. Polymer Symposium*, No. 25, p. 261-274, 1974.

13. *Rubber Age*, v. 97, No. 1, p. 78, 1965.

14. *Chemical Engineering*, v. 71, No. 11, p. 88, 1964.

15. *Journal of Spacecraft and Rockets*, v. 12, No. 10, p. 599-606, 1975.

16. *Rubber Age*, v. 95, No. 6, p. 893, 1964.

17. *Materials in Design Engineering*, v. 60, No. 7, p. 145, 1964.

18. *Chemical Abstracts*, v. 65, No. 6, p. 9148, 1966.

19. *Chemical Abstracts*, v. 67, No. 26, p. 11135, 1967.

20. Patent USA No. 3663495.

21. Patent USA No. 3300079.

22. E. Stampfl, E.M. Landsbaum. Solid motor aft closure insulation erosion. *AIAA Paper*, No. 1238, p. 11, 1973.

23. *Scientific and Technical Aerospace Reports*, v. 3, No. 6, p. 937.

24. R.A. Grippi. *Design fabrication and testing of the applications technology satellite apogee motor insulation*. Jet Propulsion Laboratory, 28 pp., 1970.

25. J.G. Sommer. Trowelable rubber rocket insulation for use at 2800'C. *Rubber Chemistry and Technology*, v. 57, No. 4, p. 843-854, 1984.

26. *Missiles and Rockets*, v. 4, No. 21, p. 26, 1964.

27. *Iron Age*, v. 193, No. 24, p. 141, 1964.

28. *Materials in Design Engineering*, v. 61, No. 5, p. 21, 1965.

29. *Missiles and Rockets*, v. 16, No. 3, p. 27, 1965.

30. *Automotive Industries*, v. 133, No. 11-12, p. 36, 1965.

31. *Missiles and Rockets*, v. 14, No. 11, p. 36, 1964.

32. Aviation Week and Space Technology, v. 80, No. 12, p. 57, 58, 61, 1964.

33. Mitrovic Ljubica, Hrabar Petar. Delovanje dvobaznih baruta na vulkanizate. *Deo III. Nauc.-tehn. pregl. VT1*, v. 27, No. 2, p. 17-27, 1977.

34. R.E. Morris, *Packageable high expansion ratio nozzles for solid propellant rocket motors.* Aerojet-General Corp., v. 1, p. 219, 1967.

35. H.L . Moody, F.C. Price. Prediction of nozzle material performance for NASA 260 SL-3 motor. *Journal of spacecraft and rockets,* v. 6, No. 3, p. 273-279, 1969.

36. *Patent USA* No. 3472812.

37. D.R. Hender. Elastomer Thermal Response in an Aerodynamic Heating Environment. *AIAA/ASME 3rd Joint Thermophysics, Fluids, Plasma and Heat Transfer Conference,* St. Lonis, Missouri, June 7-11, 1982.

38. R.A. Ellis, R.N. Hammond, P. Donguy. Advanced Space Motor Demonstration, *AIAA Paper,* No. 1270, pp. 7, 1980.

39. R.A. Ellis, R.N. Hammond, P. Donguy. AIAA ASME 16th Joint Propulsion Conference, Hartford Conn. June 30-July 2, 1980. *AIAA Paper,* No. 1267, pp. 14, 1980.

40. *AIAA Paper,* No. 1102, pp. 13, 1980.

41. *AIAA Paper,* No. 1214, pp. 7, 1979.

42. G. Grande, U. Scolastico. An Apogee boost motor for the Geos European satellite. *AIAA Paper* No. 1336, p. 10, 1975.

43. *AIAA Paper,* No. 1271, pp. 10, 1980.

44. *Aviation Week and Space Technology,* v. 95, No. 7, pp. 48-49, 51, 1971.

45. *Patent USA* No. 3562304.

46. *Composites,* v. 2, No. 3, p. 180-184, 1971.

47. *Patent USA,* No. 3301785.

48. *Aerospace Daily,* v. 125, No. 26, p. 170, 171, 1984.

49. *Iron Age,* v. 192, No. 25, p. 75, 1963.

50. *Materials in Design Engineering,* v. 51, No. 5, p. 5, 1960.

51. *Patent USA* No. 3421970, 1969.

52. *Journal of the American Cermic Society,* v. 52, No. 5, p. 178, 1969.

53. *Plastics Week,* v. 29, No. 6, p. 4, 1969.

54. *Patent USA* No. 3686868, 1972.

55. *Steel,* v. 153, No. 25, p. 33, 1963.

56. *ARS Journal,* v. 30, No. 11, p. 1099, 1960.

57. *Rubber Age,* v. 93, No. 4, p. 625, 1963.

58. A.R. Shalders. The development of elastomeric formulations to withstand extremely high temperatures and pressures. *Journal IRI,* v. 3, No. 1, p. 25-28, 1969.

59. *Missiles and Rockets,* v. 10, No. 9, p. 29, 32, 1962.

60. *Missiles and Rockets,* v. 11, No. 18, p. 23, 1962.

61. *Doc.-Airspace,* No. 107, p. 69, 1967.

62. *Patent France* No. 2303957, 1976.

63. M.G. Voronkov, V.P. Mileshkevich, Ju.A. Jujelevsky. *Siloxane Bond,* Nauka, Novosibrish, 1976, p. 23.

64. A.S. Kuz'minsky, E.A. Goldovsky. About the Influence of Molecular Oxygen on Degradation of the Main Chain of Polydimethylsiloxane Rubber, *Doklady Akademii nauk SSSR,* v. 149, No. 3, 1963.

65. V.V. Rode, M.R. Verhotin, S.R. Rafikov. Investigation of Thermal Degradation and Stabilization of Polydimethylsiloxane, *High Molecular Compounds*, v. XI, No. 7, p. 1529-1538, 1969.

66. Ju.A. Alexandrova, T.S. Nikitina, A.N. Pravednikov. *Investigations of the Mechanism of Thermal Degradation of Polydimethylsiloxanes, v.A*(X), No. 5, pp. 1078-1084, 1969.

67. M.A. Verhotin. *Thermal Degradation of Some Polyelementode Methylsiloxanes*, Dissertation on Candidate of Chemical Sciences Graduation, INEOS, 1968.

68. K.A. Andrianov, I.T. Gridunov, L.M. Hananashvili, A.S. Sergeev, Degradation of Siloxane Elastomers, *Caoutchouc i Rezina*, No. 11, pp. 19-21, 1968.

69. K.A. Andrianov, *Polymers with Inorganic Main Chains of Molecules*, Izdatel' stvo AN SSSR, pp. 62-69, 192-193, 1962.

70. C. Wipple, J. Thorme. Performance of elastomeric silicones in ablative and space environments. *Rubber Chemistry and Technology*, v. 39, No. 4, p. 1247-1257, 1966.

71. *New Caoutchoucs*. Collection of translations of papers under the edition of V.F. Evstratov and F.I. Yashunskaya, Izd. Inostrannoy Literatury, pp. 453-478, 1958.

72. N.B. Baranovskaya, M.Z. Zakharova, A.I. Mizikin, A.A. Berlin, Catalytic Hardening of Polydimethylsiloxane at Room Temperature, *Doklady AN SSSR*, v. 122, No. 4, pp. 605-608, 1958.

73. N.B. Baranovskaya, A.A. Berlin. The method of Cold Vulcanization of Polydiemthylsiloxanes and the investigation of Reaction Mechanism, Collection, Rubber, Impermeable and Hermetizing Materials, Oborongiz, pp. 5-15, 1960.

74. O.N. Dolgov, M.G. Voronkov, M.P. Grinblat. Silicon-organic Liquid Rubbers and Materials, Based on Them, Leningrad, Khimin., pp. 24-21, 1975.

75. V.V. Severnyi, R.M. Minasian, N.A. Karjutina. The Investigation of the Process of Cold Vulcanization of Siloxane Rubbers, In Collection of thesis of the reports on V-th International Conference on Chemistry and Application of Silicon-organic Composition, January 29-February 1, 1980, Tbilisi, Part I, Moscow, GNIIChTEOS, p. 310, 1980.

76. R.M. Minasian, V.V. Severnyi, Specificity of Vulcanization of Dihydroxipoly dimethylsiloxanes by Aminopropyltrietoxisylane. In *Collection of Report Thesis of VI-th All Union Conference on Chemistry and Application of Silicon-organic Compositions*, Riga, April 22-24, 1986, p. 109, 1986.

77. O.G. Rygova, B.V. Molchanov, A.V. Kissin, E.V. Denisova. *Kinetic Features of Cold Vulcanization of Dihydroxiorganosiloxanes*, as previous one, p. 62.

78. V.A. Vysotsky, M.P. Grinblat. The influence of Temperature and Humidity on Processes of Hardening of Silicon-organic Composition, Applied for Hermetization of Radioelectronic Devices, In *Collection of Report Thesis "Hermetization of Radioelectronic and Electrotechnical Devices by Polymeric"*. Materials, Discussion, May 16-17, 1989, Leningrad, LDNTP, p.79, 1989.

79. V.D. Lobkov, I.G. Kolokol'tseva, G.I. Safiantz, M.P. Grinblat. *Tendencies of the Development, Application and Conditions of Ecology of Siloxane Rubber Production, Production and Application of Elastomers,* Series 1, p.p. 24-27. CNIITEnergTekhnM, 1991.

80. *Polymer Encyclopedia,* under edition of V.A. Kargin, Sovetskaya Ensiklopedia, Moscow, v. l, p. 783, 1971.

81. *Bibliographic Pointer for Reviewing Literature on Silicon-organic Compounds,* under the editorship of V.M. Mironova, series 1 - 5, NIITEKHIM, Moscow, p. 24, 1975.

82. M. Shetz, *Silicon Rubber,* Khimia, Leningrad, p. l92, 1975.

83. A.P. Troitzkiy. *Industry of Synthetic Rubber in the USSR,* Theme review, CNIITEnefTekhKhimProm, Moscow, p.p. 25-27, 1980.

84. V.P. Mileshkevich. State and Perspectives of Investigations in the Branch of Synthesis of Siloxane Rubbers with High Thermo-Freezing and Oil Stability, *Caoutchouc i Rezina,* No. 6, p.p. 4-9, 1978.

85. M.F. Bukhina, S.K. Kurlyand, *Freezing Stability of Elastomers,* Khimia, Moscow, p.p. 151-154, 1989.

86. V.D. Lobkov, G.E. Novikova, I.G. Kolokol'tzeva, L.A. Novinskaya. Silicon-organic Rubbers with High Freezing Stability, *Caoutchouc i Rezina,* No.12, p.p. 5-7, 1980.

87. V.D. Lobkov, I.G. Kolokol'tzeva. Synthetic Rubbers with High Freezing Stability, *Caoutchouc i Rezina,* No. 1, p.p.43-45, 1981.

88. A.I. Marey, G.P. Petrova. Crystallization and Mechanical Properties of poly-dimethylsiloxanes under Conditions of Low Temperatures, *Caoutchouc i Rezina,* No. 9, p.p. 9-12, 1976.

89. E.A. Goldovskiy, L.A. Korotkih, A.A. Dontsov. Chemical Changes of Phenilic Groups of Polysiloxanes at Thermal Oxidation, *High molecular Compounds,* 22B, No. 7, p.p. 512-514, 1980.

90. E.A. Goldovskiy, A.A. Dontsov. Regularities of the Process of Heat Aging of Rubbers from Siloxane Caoutchoucs and the Ways of the Increase of Their Stability, *Caoutchouc i Rezina,* No.4, p.p. 42-46, 1980.

91. S.V. Sokolov. Thermostable elastomers, *Caoutchouc i Rezina,* No. 12, p.p. 22 23, 1987.

92. C. Whipple, Silicones for a Variety of Aerospace Applications. *VI National SAMPE Symposium Society of Aerospace Material and Process Engineering,* 1963.

93. *Industrial and Engineering Chemistry,* v. 53, No. 8, p.28a, 1961.

94. *Missiles Space,* v. 11, No. 1, p. 14, 1963.

95. W. Rigling, I. Penkacik, E. Williams. The Effects of Solid Propellant Rockets Exhaust on Elastomeric Protective Coatings. *VII National SAMPE Symposium. Adhesives and Elastomers for Environmental Extremes*, 1964.

96. *Materials in Design Engineering*, v. 58, No. 6, p. 169, 170, 1963.

97. *Product Engineering*, v. 32, No. 25, p. 9, 1961.

98. *Electronic News*, 26 June, 1961.

99. *Iron Age*, v. 189, No. 7, p. 103, 1962.

100. *Science News Letter*, v. 85, No. 5, p. 69, 1964.

101. *Adhesive Age*, v. 5, No. 1, p. 25, 1962.

102. A. Mersberg, J. Nee. Heat resistant protective coatings. *Material Performance*, v. 19, No. 12, p. 13-17, 1980.

103. *Scientific and Technical Aerospace Reports*, v. 3, No. 6, p. 937, 1965.

104. *Chemical Abstracts*, v. 65, No. 6, p. 9148, 1966.

105. *Design News*, 21/VI, v. 23, No. 13, p. 16, 1968.

106. *Chemical Engineering*. v. 71, No. 11, p. 88, 1964.

107. R. Joly. Des silicones leurs emploes dans l'industrie aeroanutique. *Techn.et sci aeronaut et spat*, No. 3, 240-252, 1976.

108. V. Hribar. A critique on Internal Insulation Materials for Solid Propellant Rocket Motors. *Journal of Spacecraft and Rockets.*, v. 3, No. 9, p. 1434, 1966.

109. J. Frederiks, M. Short. Material evaluation under direct rocket exhaust impingement. AIAA/SAE 8'th Joint Propulsion Specialist Conference. 29 XII. 1972. New Orleans. Louisiana. *Journal of Spacecraft and Rockets.*, v. 10, No. 8, p. 533-535, 1973.

110. R. Campbell, J. Ramseyer, A. Hurtress. *Development of a silicone ablator for high-heat-flux and high-shear-rate conditions*. Space Simul. Process Symposium, New York City, 1972, Washington, D.C., p. 725-748, 1972.

111. *Patent USA*. No. 4448742.

112. *Aviation Week and Space Technology*, v. 114, No. 17, p. 175, 176, 1981.

113. R. Laramee, J. Mathis, S. Cardal. *26th JANN AF Solid Propulsion Meeting, 14-16 July 1970 in Washington, DC*, Tiokol Chemical Corporation Wasatch division.

114. *Patent USA*. No. 4442666.

115. D. Butler. *Thermal protection systems based on silicone solid gum rubber with various fibers, fabrics, and additives. Materials and Processing-Move into the 90's*, edited by S. Benson, T. Cook, E. Trewin and R. Turner. Elsevier Science Publishers, B.V. Amsterdam, 1989.

116. *Patent USA*, No. 3961476.

117. *Patent USA*, No. 4042441.

118. *Interavia*, No. 9, p. 933, 934, 1988.

119. *Patent USA*, No. 4031605.

120. L. Cohen, H. Couch, T. Murrin. Performance of ablator materials in ramjet environments. *AIAA Paper*, No. 697, p. 17, 1974.

121. *Aviation Week and Space Technology*, v. 112, No. 26, p. 52, 53, 1980.

122. *Aviation Week and Space Technology*, v. 110, No. 5, p. 103, 116, 117, 133, 135-137, 139, 1979.

123. *Aviation Week and Space Technology, v. 102, No. 15, p. 40-43, 1975.*

124. *AIAA* Paper, No. 1108, p. 1-11, 1978.

125. *Patent USA*, No. 4652476.

126. B.F. Gonthier, J.M. Tanzia. *Minimum smoke rocket motors with silicone inhibitors*, Copyright. American Institute of Aeronautics and Astronautics, p.p. 11, 1984.

127. H.T. Couch. Thermochemical properties of a silicone elastomeric allator. *AIAA Paper*, No. 73-741, pp. 1-13, 1973.

General Information on Hydrodynamic and Conformational Parameters of the Macromolecules of Cellulose and Its Derivatives in Non-Aqueous Solutions

Vera V. Myasoedova, Gennady E. Zaikov
Institute of Chemical Physics of Russian Academy of Sciences

The selection of effective non-aqueous solvents for cellulose opens wide perspectives for the obtainment of fibers, films, membranes, for the synthesis of cellulose derivatives under homogeneous conditions and for the analytical control during technological processing into derivatives [1]. However, the state of cellulose in non-aqueous solutions has not been studied comprehensively enough. The data on the estimation of conformational and hydrodynamic parameters are of contradictory character [2]. The appearance of non-aqueous solvents for cellulose gives the possibility of a stricter estimation of the conformational parameters of chain and of the obtainment of the interrelation between the molecular mosses of cellulose and its derivatives and characteristic viscosity of solutions. The rigidity of the cellulose chain can be changed essentially due to the salvation with non-aqueous solvents and to the formation of polymer-solvent complex [3]. Here, depending on the specificity of the complex formed, both the increase and decrease in chain rigidity can occur.

Estimation of the polymer chain rigidity in solution is important both for the theory and practice of polymer science. In particular, the phase state and theological characteristics of system play a considerable role when obtaining fibers from spinning solutions, films, thickening agents of printing dyes.

Liquid-crystalline state allowing to obtain items with improved physico-mechanical and unique optic properties is the most favorable one. However, for many cellulose derivatives due to insufficient rigidity of the polymer macromolecules it is either impossible to obtain CC-state at all or it is achieved at very high polymer concentrations.

Such highly concentrated solutions are unsuitable for the technological processing because of extremely high viscosity. In some cases, the selection of an appropriate solvent can help to achieve the increase in macromolecular rigidity and simultaneous decrease in viscosity; this opens wide perspectives for the perfection of the production of cellulose derivatives and products of their processing.

1.1. Modeling of the Conformational Characteristics of the Macromolecules of Cellulose and Its Derivatives

The inclination to realization of thermotropic and lyotropic mesomorphism of polymers together with the chemical nature of mesogeneity is determined to a considerable extent by the degree of anisodiametry of macromolecules [4]. In this connection, the analysis of the form of macromolecules and of the influence of different physico-chemical parameters on it allows to carry out a purposeful approach to the problem of realization of the lyotropic LC-state.

The analysis of the form assumes finding of the interrelation between the parameters characterizing the size of a molecule (distance between the ends of polymer chain h, inertia radius R) and its molecular characteristics (molecular mass M and contour length L). This problem is resolved within the frame of model concepts of the mechanism of chain molecule flexibility. In the theoretical bases of the physics of polymers, free-conjugated, rotary-isomeric and persistent mechanisms of flexibility seem to the most widely used ones [5].

In the model of free-conjugated chain that as developed by Kuhn [6] and Ulark [7], the real chain is substituted by an "equivalent" one, consisting of N linear segments of length A, the spatial orientations of which are mutually independent. The condition of "equivalence" assumes the selection of N and A values (L=N*A) at which the linear dimensions of a model free-conjugated chain correspond to the dimensions of a real chain molecule. The length of the segment into which the chain is divided serves here as a measure of the equilibrium rigidity of the chain and is called Kuhn segment.

Due to internal thermal motion macromolecules in solution can assume an infinite member of "momentary" conformations [8]. Average-square distance between the chain ends $\langle h^2 \rangle$ and average-square inertia radius $\langle R^2 \rangle$ are the characteristics of the statistical dimensions of a molecule. Linear statistical dimensions of a free-conjugated chain interrelate with the length of Kuhn segment as follows:

$$\langle h^2 \rangle = L^*A; \quad \langle R^2 \rangle = (L^*A) / 6 \qquad (1.1)$$

The further development of the model concepts of the conformational state of polymer chain were favored by the attempts at improvement of the agreement between the flexibility mechanism and structural-geometric parameters of macromolecules: virtual angle values and valent long lengths. A chain consisting of an arbitrary number n of links with length l that are conjugated and rotate at $\tau-\theta$, valent angle with respect to each other, is the simplest variant of such models which were considered by Eyring [9]. For a long enough chain ($n\rightarrow\infty$):

$$\langle h^2\rangle = nl^2 (1+\cos\theta)/ (1-\cos\theta) \tag{1.2}$$

$$\langle R^2\rangle = (nl^2) / 6 (1+\cos\theta) / (1-\cos\theta) \tag{1.3}$$

It follows from correlation (1.2) and (1.3) that a long enough chain with non-hindered rotation at fixed valent angle... becomes Gaussian (i.e. equivalent to free-conjugated), since its linear dimensions grow proportionally to the contour length of chain (n-l). It follows from here that it is possible to apply the statistics of free-conjugated chains to real polymer molecules of considerable length and to use Kuhn segment length as a measure of their rigidity in solution.

The approach postulating non-deformability of valent angles and bonds, while being time for the flexible chain polymers, is inapplicable to semi-and rigid chain polymers, to which cellulose and its derivatives refer. It is common knowledge that the molecules of the polymers possessing high equilibrium rigidity, at sufficient length, have in solution the form of the Gaussian tangle [10]. The impossibility to imagine in such form the macromolecules of polypeptides [11], cellulose derivatives [12], DNA [13] which contain only a small number of Kuhn segments in the chain is connected with the fact that the assumption of deformability of valent angles (and, sometimes, bonds, especially hydrogen ones) in polymer molecules with increased equilibrium rigidity does not work [10].

Realization of the flexibility mechanism, while taking deformability into account, is possible in the model of persistent chain [14]. This model takes into account the orientational short range action of elements l which constitute the chain. Here, as distinct from the free conjugated chain and chain, the orientation of the segments of which is limited by valent angles, the spatial orientations of the neighboring elements of persistent (worm-like) chain are not mutually independent: the direction of the first element is to some extent, communicated along the chain. When going to a continuous worm-like chain (l+0), the curvature of the latter is constant along its entire length and determined by the length of persistence q. The dimensions of the worm-like chain are determined from Porod's equation [15]:

$$\langle h^2\rangle = 2q^2 (L / q-1 + \exp (-L/q) \tag{1.4}$$

For a long enough persistent chain, its dimensions obey Gaussian distribution; here persistent length equals to half-length of Kuhn segment of the Gaussian chain equivalent to it: A = 2q. For an extreme case of short worm-like chain its form approximates to stick like: $\langle h^2 \rangle = L^2$.

The model of the worm-like chain in application to the macromolecules of cellulose and its derivatives has an advantage, which, to our opinion, comes to the denial of *a priori* assumptions about the flexibility mechanism (constancy of valent angles and bond) in combination with the possibility of the description, within the frames of this model, of the entire variety of the forms which macromolecules assume is solutions (from rod-like to the statistical tangle conformation).

1.2. Hydrodynamic and Conformational Parameters of Cellulose and Its Derivatives in Non-Aqueous Solutions

Those is a series of experimental methods allowing to obtain information on the conformational structure and rigidity of a macromolecular chain. The translator and rotary friction manifesting itself at diffusion, sedimentation and viscometry of polymer solutions are the main phenomena reflecting the peculiarities of the structure of polymer molecules. Quantitative interpretation of experimental data on molecular level is possible on the basis of revealing of the interrelation between conformational characteristics of molecules with their hydrodynamic properties in solution. The polymer molecule under investigation is modeled here as a geometric body of this or that configuration, the translator and rotary motion of which is described with application of the laws of the hydromechanics of macroscopic bodies in viscous medium. The worm-like cylinder, which is understood as the cylinder curved so that the form of its axis line can be described by the equation of the worm-like chain, is the model body which reflects most adequately the hydrodynamic behavior of a molecule of rigid-chain polymer in solution [16-20].

Description of the intrinsic viscosity [η] of a solution of rigid-chain molecules on the basis of the model of the worm-like cylinder was developed by Yamakawa. On the first variant [21] calculations of [η] were carried out without taking edge-effects into account. Afterwards, taking into account the semi-spherical form of the ends of the worm-like cylinder, Yamakawa [22] obtained the following expression for the region of values L / A > 2.278:

$$M/[\eta] = \phi_\infty^{-1}(M_L/A)^{3/2} \cdot M^{1/2} \cdot \varphi(L,d,A) \quad (1.5)$$

where $M_L = M/L$, $\phi_\infty = 2.87 \cdot 10^{23}$ mol^{-1} and function φ equals:

$$\varphi = 1 - [C_1(A/L)^{1/2} + C_2(A/L) + C_3(A/L)^{3/2} + C_4(A/L)^2] \qquad (1.6)$$

Coefficients C_1, C_2, C_3 and C_4 depend on d/A and are represented by polynomial series elsewhere [21]. Introduction of a complex functional dependence of φ on L, d and A is caused by the necessity of taking into account the influence of the effects of "excess" volume and "leakage" of macromolecules on their hydrodynamic properties.

"Leakage" of the chain means equality of the resistance experienced by the chain during its motion in solvent to the sum of the resistance which would be experienced by all its elements, if each of them moved in absence of the rest. In this sense, completely extended chain possesses the greatest "leakage". When the chain molecule "rolls up" into a tangle, there occurs the screening of the segments that are situated far from its surface, which is equivalent to the decrease in hydrodynamic interaction of the molecule with solvent. The degree of "rolling up" of the chain depends on its equilibrium rigidity, length and thermodynamic quality of solvent. For the macromolecules of cellulose and its derivatives, conformation of which in the region of molecular masses ~$10^5 \div 10^6$ g-mol^{-1} is different from tangle-like one, the effect of "leakage" exerts a considerable influence on their viscometric behavior.

The effect of excluded volume takes into account the limitation of the number of mutual orientations of the chain elements due to their mutual non-overlapping, resulting in the increase of the linear dimensions of the macromolecular tangle:

$$\langle h^2 \rangle^{1/2} = \alpha \cdot \langle h^2 \rangle_\theta^{\ 1/2} \qquad (1.7)$$

where α is the degree of swelling of the tangle, and $\langle h_2 \rangle_\theta^{1/2}$ are its linear dimensions in absence of volume effects. The influence of the effect of excluded volume on the hydrodynamics of molecules becomes stronger with the increase in their length and decrease in equilibrium rigidity: while the macrochain rolls up into the tangle, the number of intramolecular contracts increases - volume effects become stronger.

Since the selection of θ-conditions (conditions under which the volume effects are compensated for by the mutual attraction between the chain elements) for a considerable number of polymers meets with great differences [23], estimation of the equilibrium rigidity of the chain, which is not disturbed by the volume effects, becomes simpler when investigating the hydrodynamic properties of low-molecular fractions of polymers. The equation of Mark-Kuhn for the non-disturbed chain:

$$[\eta]_\theta = K_\theta \cdot M^{1/2} \qquad (1.8)$$

predicts the linear increase in the intrinsic viscosity with $M^{1/2}$:

$$[\eta]_\theta/M^{1/2}=K_\theta=\phi_\infty(LA/M)^{3/2} \quad (1.9)$$

Non-ideality of solvent results in a quicker increase in the dimensions of the macromolecule with the increase in its molecular mass. Stockmayer and Fixman [24] give the correlation which takes into account the dimensions of the disturbed chain:

$$[\eta]/M^{1/2}=\phi_\infty(LA/M)^{3/2}+0.51\phi_\infty BM^{1/2} \quad (1.10)$$

Extrapolation of the dependence of experimental $[\eta]/M^{1/2}$ values on $M^{1/2}$ at condition M = 0, in correspondence with equation (1.10) allows to obtain $(A / M_L)^{3/2}$ value and determine parameters A of the chain, which is not disturbed by volume effects.

Figure 1.1 confirms the applicability of Stockmayer-Fixman method for estimation of non disturbed length of Kuhn segment: $[\eta]/M^{1/2}$ at M = 0 for polyvinylcynnamate in good solvent corresponds precisely to $([\eta], M^{1/2})\theta$, which was obtained by direct measurement in θ–solvent.

With increasing equilibrium rigidity of the chain the influence of volume effects on the macromolecular sizes becomes weaker: smaller number of intramolecular contacts corresponds to smaller degree of "rolling up". The truth of this statement is confirmed by experimental facts. In particular, in confirmation of his idea about the predominant role of the equilibrium rigidity of the chain in comparison with the role of volume effects in the formation of conformational properties [10], Tsvetkov gives the data on negative temperature coefficients of the intrinsic viscosity of the solutions of cellulose derivatives.

Application of theoretical formula (1.5) to estimate the non-disturbed dimensions of the molecules of rigid chain polymers comes across considerable difficulties: dependence of M / [η] on $(L/A)^{1/2}$ is non-monotonous. It goes through a minimum at small L, which relates to the deviation of the macromolecular conformation from the tangle-like one. With increasing L/A, the dependence represented by expression (1.5) degenerates into asymptotic straight line which with good approximation can be described by the formula below [25]:

$$M/[\eta]=\phi_\infty^{-1}(M_L/A)^{3/2}\cdot M^{1/2}+2.2\phi_\infty^{-1}(M_L^2/A)[In\,A/d)-0.755] \quad (1.11)$$

which is true for long worm-like spherocylinders (L /A>20).

Representation of experimental data in the form of $M/[\eta]=f(M^{1/2})$ dependence allows to determine the value of the length of Kuhn segment of the chain from the value of the tangent of its slope, M_L being known:

$$\partial(M/[\eta])/\partial M^{1/2} = (M_L/A)^{3/2} \cdot \phi_\infty^{-1} \quad (1.12)$$

According to the section that is cut off in the ordinate axis and using A, that is calculated from formula (1.12), it is possible to determine the value of the hydrodynamic diameters d of macromolecules.

High equilibrium rigidity of macromolecules limits the applicability of extrapolation formula (1.11) to the region of high-molecular fractions of polymers. With increasing skeleton rigidity of the chain, the deviation from the properties of Gaussian tangle becomes noticeable at large molecular masses, and, respectively, the deviation from linear dependence (1.11) is noticed at higher $M^{1/2}$. Figure 1.2 shows $M / [\eta] = f(M^{1/2})$ dependencies for the fractions on cellulose nitrate with degrees of substitution 2.4 (1st dependence) and 1.14 (2nd dependence). The nitrate with lower value of the degree of substitution possesses lower equilibrium rigidity (see Table 1.2). Noticeable deviations from linearity begin for it at $M^{1/2}<400$.

For the sample with higher equilibrium rigidity (cellulose nitrate with $\gamma=2.4$), it is impossible to obtain the linear dependence of (1.11) form over the entire region of investigated M. In such a situation it is of interest to find an analytical expression in the form of a simple linear dependence between the viscometric data and molecular weight over a wider region of L/A values. Such an expression was obtained elsewhere [26]:

$$(M^2/[\eta])^{1/3} = \phi_\infty^{-1/3}(M_L/A)^{1/2} \cdot M^{1/2} + (k/4.65A_o)(M_L/3\pi)(\ln[A/d] - 1.056) \quad (1.13)$$

where A_o is the hydrodynamic invariant for a given system polymer-solvent, k is the Boltzmann constant.

Expression (1.13) allows to describe in the form of linear dependence $(M^2 / [\eta])^{1/3} = f(M^{1/2})$ over a wide range of molecular weights. The correctness of this is illustrated by dependence 3 in Figure 1.2. From the slope of the straight line (1.13) one can determine the parameter of the equilibrium rigidity of macrochain A:

$$\partial(M^2/[\eta])^{1/3} / \partial M^{1/2} = (M_L/A)^{1/2} \cdot \phi_\infty^{-1/3} \quad (1.14)$$

at known M_L and $\phi_\infty=2.87 \cdot 10^{23}$ mol^{-1}. The section, that is cut off by the straight line (1.13) in the ordinate axis, allows to calculate diameter of chain d.

These considered methods of the asymptotic approximation to the theoretical dependence of intrinsic viscosity of the worm like spherocylinders (1.5) have a considerable disadvantage coming to the usage, in the calculations, of the value of the length of a projection of monomeric link onto the direction of chain $\lambda = Mo/M_L$; this value was obtained during the analysis of its chemical structure, or from the data of X-ray structural analysis of the polymer in condensed state. However, identification of these data with the geometric properties a molecular chain in solution is not always admitted, especially in the cases when the formation of a specific secondary structure within a molecule in solution is possible. Experimental determination of λ value is possible only for the polymers with very high chain rigidity, conformation of which in really used region of molecular masses is close to stick-like. Nitrates, for which λ is 5.1±0.1Å [10,21], meet this requirement to the greatest extent among the cellulose derivatives. This value, which is given here for nitrates, is also used for the other cellulose derivatives [27,28] in practical calculations. Determination of A or M_L, using expressions (1.11) and (1.13) which describe the asymptotic approximation of dependence (1.5), is possible at independent experimental determination of A or d.

This possibility is realized in the method of Bohdanecky [29], which is based on the assumption of the equality between the hydrodynamic volume occupied by 1g of the worm-like chain and partial specific volume of polymer v; using a series of empirical correlation (formulas (11-13) in [28]), this allow to calculate d. The slope and intersection of $(M^2/[\eta])^{1/3} = f(M^{1/2})$ dependence with the ordinate axis allows to calculate A and M_L, d being known.

For example, let us consider the determination of the hydrodynamic and conformational parameters in the non-aqueous solution of cellulose in the mixtures of trifluorocetic acid with chlorinated hydrocarbons [1].

In order to establish the interrelation "molecular mass intrinsic viscosity", the authors of [30] experimentally determined the molecular mass characteristics of the cellulose samples with different degrees of polymerization and the intrinsic viscosity of non-aqueous solutions in the mixtures of trifluoroacetic acid (CF_3COOH) with chloroform ($CHCl_3$), methylene chloride (CH_2Cl_2), 1.2-dichlorethane ($C_2H_4Cl_2$). The measurements were carried out for the diluted solutions of cellulose with the degrees of polymerization from 150 to 910 using the methods of sedimentation and viscometer. The sedimentation analysis of the non-aqueous solutions of cellulose in CF_3COOH mixtures with $CHCl_3$, CH2Cl2 and $C_2H_4Cl_2$ was carried out using the method of unfixed equilibrium [31] on analytical ultracentrifuge 3180 ("MOM", Hungary) at rotor temperature 298±0.1 grad. With residual vacuum 1.3Pa.

The concentration of cellulose in solution was varied from 0.08 to 0.25 g/dl. The numerical processing was performed on the basis of "Aspect-2000" computer according to [32]. The viscosity was determined on the viscometer with "suspended" level at temperature 298+0.1 grad. for a series of solutions within the same concentration interval. The value of the intrinsic viscosity [η] was determined by double logarithmic extrapolation at zero of η_{yg} / c and $\ln \eta_{oth} / c$ dependencies.

Non-destructuring solvents were selected in order to determine reliable molecular mass characteristics of cotton and wood cellulose. The optimum ratios of components in CF_3COOH mixtures with $CHCl_3$, CH_2Cl_2 and $C_2H_4Cl_2$ were selected taking into account the comparative analysis of the data on cellulose solubility and acid self-association in non-aqueous solvents [33] and equal 0.3 mol. fr. chlorinated hydrocarbons in the mixture with acid.

The results of the sedimentation analysis of cellulose solutions in CF_3COOH and its mixtures with chlorinated hydrocarbons testify to a considerable influence of the nature and composition of solvents on the form of sedimentograms.

Figure 1.3 show the curves of the sedimentation analysis of 0.1% mass solutions of cellulose in CF_3COOH and its mixtures with CH_2Cl_2, $CHCl_3$ and $C_2H_4Cl_9$. In the case of cellulose solutions in CF_3COOH + 0.3 m.f.$C_2H_4Cl_2$ and CF_3COOH + 0.3 m.f. $CHCl_3$, the form of sedimentograms is traditional for diluted solutions of polymers. The forms of sedimentograms of the solutions in CF_3COOH and CF_3COOH-CH_2Cl_2 are of multimodel character and their reproduction is difficult. The average molecular mass was calculated by extrapolation at infinite dilution of the effective molecular mass using expression [33]:

$$(1/M)_{app} = 1/\overline{M}_w + A_2C + \ldots \quad (1.15)$$

The values of M_W were corrected for polydispersion, taking into account the polydispersion coefficient according to the equation below:

$$\overline{M}_w^{kopp} = \overline{M}_w(a+1)\Gamma(a+1)/K \quad (1.16)$$

where a is the exponent of Mark-Kuhn-Howink equation, Γ is the gamma function; K=2÷7..

The specific partial volume V, that is necessary for the calculation of the average molecular mass from the sedimentation data, was determined picometrically.

The values of V and ρ at 298 were 0.586 cm^3/g and 1.393 g/cm^3, respectively.

Table 1.1 carries the values of molecular mass and intrinsic viscosity.

Table 1.1 The values of molecular masses and intrinsic viscosity of cellulose samples

DP of cellulose	$M \cdot 10^{-3}$	[η]dl/g	$A \cdot 10^4$, $cm^3/g \cdot mol$
150	21.0	1.86	36
150	24.3*		
600	91.0	7.84	31
600	97.2*		
755	122.0*		
750	140.0	8.81	27
910	147.0*		
910	165.0	9.81	24

m* - values of masses of cellulose samples as shown on labels.

While the viscosity of the cellulose solutions in CF_3COOH decreases essentially due to destruction [35], the control of the stability of cellulose (DP 600 and 900) solutions in CF_3COOH-$C_2H_4Cl_2$ mixture showed that the viscous properties of the solutions (within the limits of determination errors) did not change during 2 months. This testifies to the absence of destruction of cellulose.

The measurements and calculations performed helped to build $[\eta]=f(M_w)$ and $(M^2/[\eta])=f(M^{1/2})$ dependencies and determine the constants of Mark-Kuhn-Howink equation using the least square method.

The model of persistence chain was used to estimate non-disturbed dimensions of cellulose macromolecules in non-aqueous solvent. The dependence in Figure 1.4δ, can be described well by equation $(M^2/[\eta])^{1/3}=f(M^{1/2})$.

The section that is cut off in ordinate axis $A_\eta = 34$ and tangent slope of straight line $B_\eta=0.65$, were used to calculate the values of Kuhn segment(A), diameter of macromolecule (d) and other conformational parameters in correspondence with the equations below [28]:

$$d_r / A_o = 0.5 B_\eta^4 \cdot \overline{V} / A_\eta \tag{1.17}$$

$$\lg d_r = -0.173 + \lg(d_r^2 / A_o) \tag{1.18}$$

$$A_o = 0.46 - 0.53 \lg d_r \tag{1.19}$$

$$M_L = A_\eta \phi^{1/3} / A_o \tag{1.20}$$

$$h^2 / M = (B_\eta \phi_o^{1/3})^{-2}; A = h^2 \cdot M_L / M \tag{1.21}$$

The value of conformational parameters are as follows:

$$A = 170 \cdot 10^{-8}\ \text{cm},\ d = 5.5 \cdot 10^{-8}\ \text{cm}.$$

Closeness of parameter a to unity, value of Kuhn segment and absolute values of second virial coefficients, that are given in Table 1.1 allow to conclude that cellulose in this solution is a semi-rigid-chain polymer.

A reiterating structural unit of cellulose is shown in Figure 1.5. The contour length λ of one link of cellulose, i.e. the projection of the lengths of bonds onto the chain axis, equals 5.33Å; molecular weight of link M_o= 162.

The method of calculation of A was proposed by Parkov [36] who identified concentration C, at which a sharp increase in $\eta_{yg}/c=f(c)$ dependence (see Figure 1.6), with the concentration at which the distances comparable the centers of inertia of macromolecules become comparable with the sizes of the macromolecules themselves.

This supposition allows to estimate according to C*, the geometrical dimensions of the diameter of rotation spheres of macromolecules

$$d \approx (2.05M / c^*)^{1/3} \cdot 10 \qquad (1.22)$$

where d is the diameter of rotation sphere (Å); C* is the upper concentration boundary of diluted solution (g/100 cm^3).

For rigid-chain polymers Parkov makes the assumption: $d \approx \langle h^2 \rangle^{1/2}$, which using Porod's formula (1.4) allows to determine the equilibrium rigidity of macromolecule A= 2q. Further representation of experimental [η] values in the form of dependencies (1.11) or (1.13) allows to calculate d and M_L.

The development of computational equipment allowed to carry out the comparison of experimental and theoretical calculated values of the intrinsic viscosity with functional dependence of Yamakawa (1.5) connecting [η] with the molecular characteristics of polymer (M_L, A, d). This approach to the calculation of molecular parameters allows to eliminate the assumptions that were made when theoretical dependence (1.5) was substituted by approximated expressions (1.11) and (1.13). The essence of the approach comes to variation of the values of molecular parameters (M_L, A, d, L) that are introduced into expression (1.5) and to comparison of the theoretical values $[\eta]^{Teop}$ thus obtained with experimental ones. It is considered here that a combination of M_L, A and d parameters, ensuring $[\eta]^{\varepsilon kcn} = [\eta]^{trop}$ condition, corresponds to the conformational state of the macromolecule.

Bianchi et al [27] who studied the equilibrium rigidity of cellulose in N, N dimethylacetamide-LiCl mixture, assumed the length of the projection of monomeric link of glucapyranose chain as 5.14Å.

Variation of the value of hydrodynamic diameter from 4.5 to 6 Å and of the persistent length from 70 to 140 Å showed that the worm-like cylinder with q = 110±10Å was the best to model the behavior of real cellulose macromolecules in solution (Figure 1.7).

Analogous procedures for calculating the conformational parameters were performed elsewhere [18,38], and the authors of [37] extended the number of variable parameters and included the contour lengths of chains. A value thus obtained agree satisfactory with the data are obtained by extrapolation of $(M^2/[\eta])J^{1/3}=f(M^{1/2})$ dependencies at M = 0.

Limited applicability of the variation approach to calculations of A and d conformational parameters relates to the impossibility of their independent obtainment when resolving theoretical expression (1.5): numerous pairs of d - A values correspond to $[\eta]^{\varepsilon kcn}=[\eta]^{trop}$ condition. Independent determination of d and A [36] is connected with not strict enough assumptions, that is why d and A were varied within reasonable limits. Due to imperfection of the variation approach to the determination of equilibrium rigidity of macromolecules, most investigators [16,20,39] calculated A using approximated expressions (1.11) and (1.13) of the theoretical equation of Yamakawa.

1.3. Influence of Molecular Characteristics of Cellulose and Its Derivatives on Equilibrium Rigidity in Solution

Table 1.2 carries the values of diameters and lengths of Kuhn segments of the macromolecules of cellulose and its derivatives that were calculated elsewhere [40-70] using the methods described in section 2.

It is noteworthy that the data given for one and the same polymer but by different authors differ from each other rather noticeably. For example, the values of the parameter of equilibrium rigidity for cellulose in cadoxene vary from 76 to 142 Å [40-44]. And if one takes into account that A values for different cellulose derivatives are hardly comparable with each other due to the differences in the degree of substitution γ and in the nature of the solvent used, he can understand the difficulties arising with the attempt at analyzing the influence of the type of the substituent, which is introduced into the cellulose chain, of the degree of substitution, molecular mass, solvent nature on the equilibrium rigidity of the macromolecules of cellulose derivatives.

The only fact, which is testified to by the data in Table 1.2, is that the macromolecules of the cellulose derivatives possess a heightened equilibrium rigidity. The values of Kuhn segment lengths that are characteristic of them (100 ÷ 300 Å) are by one order higher that A values that are typical of flexible-chain polymers. The flexibility of the molecular chain of cellulose and its derivatives is determined by the possibility of more or less hindered rotation of pyranose cycles around two bridge bonds OC_1 and OC_4 (Figure 1.8). At the same time, the parameter of equilibrium rigidity of the cellulose derivatives exceeds considerably the Kuhn segment length at non-hindered internal rotation A_f= 11Å [10]. This fact shows that the interaction of side groups of polyglucoside chain which prevents from the rotation around valent bonds of the main chain, does not come to usual steric disturbances only. Evidently, an essential role in these interactions is played by the hydrogen bonds which close a molecular chain into a cycle. According to the data of IR-spectroscopy [71], additional cyclization of the cellulose molecules in crystalline state is due to the formation of intramolecular hydrogen bonds (O'_3O_5) and ($O_2O'_6$) (see Figure 1.8). The determining influence of hydrogen bond (O' O) on the conformational state of the macromolecules of cellulose acetates in solution was taken by Shakhparonov, Zakurdayeva and Podgorodetsky [72] as the basis for their analysis.

Some interpretations [40,44] of experimental facts also come to the mechanism of rupture formation of intramolecular H-bonds. So, Lybina et al. [40] explain the increase in the cellulose rigidity in cadoxene, while water is being added into it, by the reduction of hydrogen bonds within a macromolecule, which were predestroyed by cadoxene.

Experimental data in Figure 1.9, shown as the dependence of equilibrium rigidity parameter A on the degree of substitution γ in the cellulose derivatives, testify to the extension of the possibilities of variation of the glucopyranose chain flexibility by way of the selection of an appropriate substituent and solvent.

Proceeding from general considerations, one can suppose that introduction of the substituents into a cellulose molecule creates additional steric disturbances to the neutral rotation of pyranose cycles, which can result in the increase of the chain equilibrium rigidity. However, if we consider the behavior of cellulose carbanilate (R=-CO-NH-C_6H_5) and nitrate (R=-C_2H_5) with similar degrees of substitution (γ=2.3). In ethyl acetate, we shall mark out that their equilibrium rigidities do not differ within the determination error limits (154, 175 and 160Å, respectively [47, 54, 58]). Though there are data on the increase in the rigidity of the main chain of comb-like polymers with increasing dimensions of side mesogen groups [10], the determining role in the mechanism of flexibility of the cellulose derivative chains is, evidently, played not by the steric effects, relating to the enlargement of radicals, but by the effects of secondary cyclization of its structure by intramolecular hydrogen bonds.

Probably, introduction of substituents into the glucopyranose chain leads to the change in localization of H-bonds within the chain: the old ones are destroyed, and the new-formed. Formation of new H-bonds is possible both between newly introduced radicals (carbanilate) and between the radicals and residual hydroxyl of the pyranose cycles (cellulose acetate). In the former case, the increase in the degree of substitution causes the increase in the chain skeleton rigidity. Burchard [12, 62] proposed the scheme of the secondary cyclization of cellulose Carbanglate by the hydrogen bond of >NH→O=C< type to explain high equilibrium rigidity of its chain (A=338 Å for carbanylate in DM80 [68]. In the latter case, the increase in the degree of substitution is accompanied by extreme change in the chain rigidity. Introduction of acetyl groups into a cellulose macromolecule leads to the increase in the rigidity of its chain due to the formation of -OH→O=C< hydrogen bonds. The number of newly formed hydrogen bonds between acetyl and hydroxyl groups is maximum at certain $\gamma^* < 3$, to which the highest value of the chain equilibrium rigidity among acetates corresponds. The magnitude of γ^* depends considerably on the evenness of distribution of the substituents within the chain. According to the data of Shakhparonov et al. [72] for the cellulose acetates in dichlormethane, the maximum rigidity is achieved at $\gamma^* = 2.90$ $[\eta]$(at $\gamma = 2.70$ $[\eta] = 285$ cm^3/g; at $\gamma^* = 2.90$ $[\eta] = 350$ cm^3/g; at $\gamma = 3.0$ $[\eta] = 275$ cm^3/g). The method of introduction of methanol and change in the composition of mixed solvent on the basis of dichlormethane was used [72] for the qualitative estimation of the number of intramolecular H-bonds. Introduction of methanol led to the rupture of (O'_3O_5) bond and conditioned the decrease in registered dimensions of the macromolecular tangles. Such an effect for cellulose triacetate solutions was not found, which, according to the authors, pointed at the absence of cyclization of the molecules of cellulose triacetate by intramolecular H-bonds. The data, that were obtained by Sato [42] when studying the hydrodynamic behavior of low substituted acetates in dimethylacetamide, testify to a double increase in their skeleton rigidity while the degree of substitution increases from 0.49 to 2.92 (Figure 1.10). Unfortunately, the lack for experimental A values for the cellulose acetates with the degrees of substitution >2, which were obtained in one solvent lays obstacles to a more detailed analysis of the influence of the degree of substitution on the chain rigidity.

Figure 1.10 shows the dependence of the equilibrium rigidity of the molecules of cellulose nitrates on the degree of substitution in several solvents. A considerable increase in A (more than 200%) upon transition to double substituted cellulose nitrates attracts the attention. The question of strong dependence of the cellulose nitrate conformation on its degree of substitution was considered by Pearson and Moore [73], however, the reason remained unknown. Estimation of the regularities of the influence of solvent on the skeleton rigidity of the cellulose nitrate molecules is complicated because of considerable differences in the data that were obtained by different authors under similar determination conditions according to [53], for nitrate with $\gamma = 2.9$ in acetone A = 332 Å; and according to [42] for nitrate with $\gamma = 2.91$ in acetone A = 260 Å. Nevertheless, it can be considered that the rigidity of the cellulose nitrate chain, on the average, increases insignificantly over the following series of solvents: acetone < ethylacetate < methylethylketone < dimethyl acetamide + 6% LiCl (Figure 1.10).

Solvent influence on the macrochain conformation was analyzed in greater detail by Kamide and Saito [74]. They mark out a considerable increase in the rigidity of cellulose diacetate ($\gamma = 2.46$) chain upon transition to the solutions in more polar solvents (Figure 1.11). One can notice a decrease in the equilibrium rigidity of polyelectrolytes (Na-salt of carboxymethyl cellulose) with increasing ionic power of solvent [42,70]: for NaCMC in the aqueous solution of sodium chloride A = 254 Å at J= 0.1 mol/l; A = 190 Å at J = 0.34 mol/l; A = 100 Å at J →∞. If one takes into account that the increase in the ionic power of the aqueous solution of sodium chloride leads to the decrease in its dielectric permittivity, then the data obtained elsewhere [70], agree qualitatively with Kamide and Saito s conclusion.

2. Peculiarities of the Rheologic Behavior of Lyotropic LC-Systems on the Basis of Cellulose Derivatives

2.1. Polymer concentration influence on the conformational state of macromolecules in solution and the rheologic characteristics

Absence of intermolecular interactions in diluted solutions was the basis condition for the analysis of the conformational state of macromolecules in sections 1.2 and 1.3. At the same time, the problem of the prediction of isotropic-anisotropic phase transitions in concentrated solutions puts forward the task of a detailed consideration of the conformational changes relating to the increase in intermolecular interaction upon transition to concentrated solutions.

The problem mentioned is resolved better for flexible-chain polymers. Let us consider the basic regularities obtained from the theoretical and experimental investigations of the solutions of flexible-chain polymers, which, as we think are also true for the rigid polymers, to which cellulose and derivatives refer. According to the data of scaling analysis [75] the transition to the concentrated solutions is conducted through rather a wide region of semi-diluted solutions with the concentration boundaries C* and C**. When $C = [\eta]^{-1}$ is achieved, partial overlap of the macromolecular tangles begins. A decrease in the statistical dimensions of molecules (эcompression) and increase in the number of intermolecular contact occurs with increasing concentration in C* < C < C** region. эCompression of the molecules comes to a stage at which their linear diminution correspond to the dimensions that are assumed under θ-conditions. The concentration, at which θ - conditions of the conformational state of molecules C** are realized corresponds to the upper boundary of the existence of semi-diluted solution. Further concentration increase is accompanied by a quick increase in the number of intermolecular contacts; the fluctuation net of bonds being formed here; this net determines the entire complex of the physico-chemical properties of concentrated solution.

The degree of эcompression of macromolecules is determined by the thermodynamic quality of the solvent used. Figure 2.1 shows the diagram of state of the solutions of a flexible polymer, which was built using the data of theoretical calculation [76].

The essential conclusion, which follows from the analysis of the diagram, is in the prediction of a considerable decrease in the skeleton rigidity of macromolecules upon transition to concentration solutions: the solvents that are used to obtain the solutions of cellulose derivatives are thermodynamically "good", i.e. $\tau=(T-\theta)/\theta \rightarrow 0$.

The hydrodynamic estimation of the disturbing action of the density of intermolecular contact net on the conformation of molecules [77] confirms qualitatively the results of the theoretical investigations: the volume occupied by polystyrene molecules in bromoform decreases with increasing concentration of polymer, and this decrease is faster if its molecular weight is higher (Figure 2.2).

Information on the conformational changes of the rigid polymermacromolecules comes only to the data obtained from computer experiments [78-80]. Birshtein notices a decrease in the extension of the branches and skeleton of the comb of comb-like polymer with increasing concentration behind the overlap edge C*. There is a conclusion [79] about the decrease in radial extension of the rays of a star-like polymer with its increasing concentration in a semi-diluted solution. The stars begin to penetrate into each other only after they assume Gaussian dimensions.

The authors of [80], who used the method of computer experiment to model the behaviors of cellulose nitrate macromolecules in thermal solution (=0), obtained, in semi-diluted and concentrated solutions, the dimensions of macromolecules that are close to Gaussian $\langle R^2 \rangle = \langle R^2 \rangle_\theta$.

Absence of experimental works in this area makes take into consideration the data of computer simulations which testify to the decrease in equilibrium rigidity of rigid polymer molecules upon transition from diluted to concentrated solutions.

2.2. REGULARITIES OF SHEAR DEFORMATION INFLUENCE ON THE RHEOLOGIC PROPERTIES AND CONFORMATIONAL PARAMETERS OF THE CONCENTRATED SOLUTIONS OF RIGID POLYMERS

The analysis of the hydrodynamic properties of diluted polymer solution 1.2 did not touch upon the influence of shear deformation on the conformational state of macromolecules. At the same, application of external mechanical field causes deformation of the macromolecules with noticeable kinetic flexibility.

The kinetic flexibility characterized the rate at which there occurs a change in the chain conformation which is accompanied by the overcome of potential barrier U_0[2]:

$$t \sim \exp(U_o / kT) \tag{2.1}$$

Application of the external mechanical field with intensity is τ, according to Eyring [82], equal to the decrease in the potential barrier in the direction of force action; this makes the conformational transitions, stretching (unrolling) the molecular chain in the direction of force action, more probable. The unrolling of the molecular tangle becomes more noticeable with growing kinetic flexibility of polymer and intensity of the field applied. The increase in tangle dimensions strengthens the hydrodynamic interaction of a molecule with the surrounding solvent; which results in additional dissipation of energy by the system. At the same time, orientation, in the direction of the stream, of the deformed particle, anisodiametry of which grows with increasing shear velocity, is accompanied by the decrease in the viscosity coefficient of the system. Of the two effects noted (the first of which - increase in the linear dimensions of a chain molecule upon deformation - leads to increasing viscosity of a solution, and the second - orientation of deformed particle in the stream - to the viscosity decrease) the second one assumes a predominant importance with increasing shear velocity; this leads to the deviation from Newton's law:

$$\tau_{\chi} = \eta \frac{\partial V_x}{\partial_y} = \eta \cdot \dot{\gamma} \tag{2.2}$$

towards the decrease in the viscosity of a solution.

The quantitative analysis of the influence of purely hydrodynamic factors (deformation and orientation of particles in stream) on the viscosity of polymer solutions was carried out by Bueche [81]. The hydrodynamic factors considered play the determining role in the mechanism of viscosity lowering with deformation increase only in diluted solutions. In semi-diluted and concentrated solutions of polymers, according to Rebinder [82], the viscous properties of a system are determined by the destroy of forming fluctuation net of intermolecular contacts. And only after is destroy the contribution from the hydrodynamic factors becomes noticeable again. This is manifested, because of the weakness of these factors in comparison with the factor of the destroy of overmolecular organization, in the existence of the second section of the independence of viscosity ($\eta \to \eta_\infty$ at $\tau \to \infty$). The sections of "quasineutonic" behavior at high values of tangent tensions were found experimentally in the curves of flow of cellulose nitrate solutions [83]. There is a structural branch between the Newtonean and quasinewtonic sections of the curve of flow; development of anomalous viscosity within the limits of this branch is connected, according to Rebinder, with the degree of destroy of the intermolecular contact net:

$$\alpha = (\eta_o - \eta)/(\eta_o - \eta_\infty) \tag{2.3}$$

The mechanism of viscosity lowering with destruction of the spatial net of bonds was considered more deeply by Graessley [84]. He introduced the concept of the induction period, which is necessary for the formation of contacts between hovered macromolecules. The duration of contacts is determined by shear velocity, the increase in which lowers the probability of the formation of intermolecular bonds. The attempt at taking into account the effect of molecular deformation, the role of which grow with progressing destruction of the bond net, was made by Williams [85] but did not prove to be successful enough: his theory predicts a considerably more intensive decrease in viscosity in comparison with experimentally observed one [86]. Such a divergence can be caused by the fact that the possibility of the formation of secondary fluctuation net by deformed molecules was not taken into account. It is marked out [87] that in the systems, where macromolecules can interact selectively, the deformation makes intermolecular interaction stronger, and the increase in the number of contacts becomes a dominating factor in comparison with the hydrodynamic effects.

The complexity taking into account of the contribution which is made into the mechanism of solution flow by each of the factors explains, to our mind, the absence of a united molecular theory of flow of polymer solutions, which would agree satisfactorily with experimental facts. Nevertheless, the qualitative analysis of the influence of molecular characteristics of polymer (molecular mass) on the rheologic properties of its solutions is possible.

The increase in polymer molecular mass which is accompanied by condensation of the net of intermolecular contact ensures the predominant role in the mechanism of polymer flow to the factor of the destruction of fluctuation net. Figure 2.3 illustrates the increase in the degree of deviation from Newtonian-behavior with increasing molecular mass of polybutadiene (curves 2 and 3). In the case, when the macrochain length is insufficient for the formation of the effective net of bonds, the change in η with τ increasing is determined by the hydrodynamic factors, the relative weakness of the action of which on the viscosity of the system leads to the preservation of the Newtonian character of flow in a considerable region of tensions.

This consideration agrees with a well-known experimental fact of the strengthening of the dependence of viscosity on molecular mass none a certain value Mcr is achieved [88-90]. Mcr divides the power dependence

$$\eta_o = K \cdot M^{\alpha} \tag{2.4}$$

into two sections with different magnitudes of power index α. For the initial section, different authors give α_1 values between 1 and 1.6 [89,90]. For the second section α_2=3.4÷3.65

The transition to the hydrodynamic regime of flow of the concentrated solutions at $\tau \rightarrow \infty$ lead to the subordination of the dependence of viscosity on molecular weight to a united law η = K·M over the entire interval of polymer molecular masses.

The increase in the number of intermolecular contacts with increasing polymer concentration in solution determines the rheologic behavior of solutions: noticeable derivations from the character of flow with increasing concentration are achieved at lower shear velocities (Figure 2.4).

Since the deformation of a particle in the stream and the number of the intermolecular contacts, that are formed by this particle, depend on the kinetic and equilibrium rigidity of the macromolecule, then the analysis of the rheologic behavior of polymer solutions from the positions of macrochain rigidity would be more consistent. However, this question was not touched upon in the literature which is accessible to us.

2.3. Viscous properties of anisotropic solutions of cellulose and its derivatives

In section 2.2 we considered the deforming and orienting section on macromolecules of the mechanical field applied to a polymer solution. The transfer of the oriented state of macromolecules from solution into fibers favors the increase in their mechanical solidity [91,92]. However, relaxation of Rensions, occurring when extrudate goes out of filler canals, a well as at the stage its coagulation, is accompanied by deorientation of macromolecules in the fiber formed; this influences significantly the solidity characteristics of the fiber [93]. In this connection of great interest is the possibility of spinning of fibers from the solutions the high degree of orientation of the molecules in which is determined not only and mainly not by the application of shear deformation but achieved spontaneously due to high equilibrium rigidity of the polymer chain.

The possibility of a spontaneous transition to highly ordered state in concentrated solutions of semi-and rigid polymers which is accompanied by the increase in the statistical dimensions of macromolecules is confirmed in several theoretical [94-97] and experimental [105-109] works. Computer simulation [80] of the behavior of cellulose nitrate in an thermal solution revealed the increase in the macromolecular dimensions with excess of the volume fraction of polymer in solution V = 0.1; in V = 0.05 ± 0.26 region-approximately by 12%; at V = 0.3 the linear dimensions of macromolecule are only by 30% less that the contour length of chain (Figure 2.6a). Theory of the average field of nematic-isotropic phase transition [94] also presupposes the spontaneous stretching of macromolecular tangles leading to the increase in their dimension (by an order at DP=100 q / L = 0.1 and energy of the interaction between monomeric links of neighboring macromolecules equaling to 0.4 ccal/mol) when the concentration-temperature boundaries of the phase transition is achieved (Figure 2.6b).

There is a series of monographs and reviews characterizing the rheologic behavior of lyotropic C-polymers [102, 112]. Generalizing the peculiarities of viscometric behavior of spontaneously ordered systems one should first of all pay attention on a considerable difference in the values of the viscosity coefficients corresponding to longitudinal and diametrical orientations of anisodiametric particle in stream [96]. This determines a noticeable decrease in the viscosity of an ordered system of such particles if they are oriented along the stream. At the same time the spontaneous ordering of macromolecules the reason for which is their heightened equilibrium rigidity insolution is masked in the systems deformed by the unrolling and orienting action of the mechanical field. This is manifested in the monotonous viscosity increase with increasing polymer concentration on solution at large enough shear velocities (curves 4 and 5 in Figure 2.7). The situation changes sharply with increasing intensity of deformation dominating influence on the ordering of particles in this case is exerted by their disposition to the formation of mesophase in solution which results in the appearance of an extreme in the concentration dependence of viscosity (curves 1-3 in Figure 2.7). The appearance of the extreme is caused by different degrees of disordering of the systems with increasing intensity of deformation in the solutions where polymer concentration is sufficient for the formation of mesophase the mutual orientation of macromolecules is preserved within the limits of domain structures at relaxation of the tension which is accompanied by comparatively small increase in the solution viscosity.

Qualitatively the same result (unrolling of macromolecular tangles and their mutual orientation) which is obtained for kinetically flexible macromolecules by way of application of external shear field and for the molecules with high equilibrium rigidity - by way of heightening of their concentration in solution made some investigators [98,99] try to induce C-state in isotropic systems by applying the mechanical field. The theoretical analysis of such possibility was carried out elsewhere [100-101]. However as noted by Kulichikhin [102], there are no experimental confirmations to the induction of C-ordering in the systems that are not disposed to lyotropic mesomorphism. As for the solutions of polymers the action of deformation in some cases makes the transition of a system into highly oriented state easier [93]. We believe that the investigators did not underline the possibility of the interrelation between the "pliability" of the temperature-concentration boundary of the phase transition and the kinetic flexibility of macromolecules. Kinetic and equilibrium rigidities do not obligatory change symbatically [2, 103] and for the molecules with high equilibrium and low kinetic rigidity the deformation should favor the realization of C-state. Maybe the concept of the kinetic flexibility will allow to explain why for some rigid polymers (polyterephthalamide n-aminobenzhydrazide [93]) the increase in shear velocity causes a noticeable displacement of the viscosity extreme into the region of lower polymer concentration (Figure 2.8) and for the others (polybenzyl-L-glutamate [96] this is not observed (Figure 2.7). The complexity of such questions is redoubled by the problematicity of identification of the concentration boundary of isotropic-anisotropic phase transition with the concentration at which the maximum viscosity is achieved. There is a lively discussion on this problem some authors speak about the identity of both concentrations (Gilbert and Patton [104]), the others consider that the phase transition occurs at the concentration that is lower than that at which the extreme value of viscosity is achieved in the concentration dependence [106]. S. Suto [107] refers to the latter and says that a more precise estimation of the concentration-temperature conditions of the formation of anisotropic phase in solution one should orient on the position of the maximum of the temperature dependence of the solution viscosity. It is noteworthy that the increase in the intensity of deformation of cellulose solutions in the mixture methyl morpholene N-oxide water [92] and of cellulose triacetate in trifluoroacetic acid[108] does not have any noticeable influence on the position of the maximum of the temperature dependence of viscosity (Figure 2.9).

Small number of the works devoted the study of the shear deformation influence on the viscous properties of the systems that are inclined for lyotropic mesomorphism as well as problematicity of correspondence between the extreme values of concentration and temperature dependencies of viscosity and the concentration-temperature boundaries of the phase transitions are the reasons for a complicated estimation of the contributions from the equilibrium and kinetic rigidities into the formation of lyotropic C-phase. At the same time analysis of the rheologic peculiarities of systems from the viewpoint of equilibrium rigidity of macromolecules seems rather logic. It is easily seen when considering the temperature influence on the skeleton rigidity of polyglucoside chain and on the position of the maximum value of viscosity in the concentration dependence for a series of the cellulose derivatives. As K is marked out elsewhere [28] the temperature increase from 25° to 150°C causes the decrease in the parameter of equilibrium rigidity of acetoxypropylcellulose chain from 150 to 58 Å. Such a tendency was marked out by the authors of [64] when studying the equilibrium rigidity of cellulose triacetate in acetic acid (at 25°C A = 150Å, and at 60°C A = 90 Å). The decrease in the chain skeleton rigidity with increasing temperature is the most widely appeared argument when explaining the displacement of the position of maximum in the concentration dependence of viscosity in the region of higher polymer concentrations [64,94]. In particular for solutions of TAC in trifluoroacetic acid - dichlormethane mixture Patel and Gilbert [109] marked out a 50% - increase in the concentration corresponding to the maximum value of viscosity while the temperature rises from 13 to 30°C. Such a regularity is common for all semi- and rigid-chain polymers forming anisotropic phase in concentrated solutions.

Beside the shear deformation and temperature the molecular mass of polymer also influences considerably the position of extremes in the concentration dependence of viscosity: with increasing molecular mass the maximum of viscosity is achieved at lower polymer concentrations (see Figure 2.7). Interpretation of this regularity from the viewpoint of the influence of equilibrium rigidity on the anisodiametry of molecules requires to assume that the skeleton rigidity of chain grows with increasing molecular mass. The question of the influence of the molecular mass of polyglucoside chain on the parameter of its equilibrium rigidity is touched upon by Ozolinsh Karkla [69], and though the authors do not concentrate their attention on this moment the data obtained by them testify to a considerable increase (~70%) in Na CMC chain rigidity in the region of the low molecular fractions of polymer (Figure 2.10).

The dependence of Kuhn segment value on the molecular mass of cellulose derivatives is also noticed by Saito [42] and beside in some cases the increase in molecular mass is found to cause the decrease in the skeleton rigidity of polyglucoside chain (cellulose acetate in acetone) and in the other cases - its increase (cellulose nitrate in acetone). An indirect proof of the decrease in the equilibrium rigidity of the cellulose chain with the increase in its contour length is obtained elsewhere [92]: the authors marked out the contraction of the temperature boundaries of the existence of anisotropic state of cellulose in the aqueous mixtures of N-methylmorpholene-N-oxide while the degree of polymerization increases from 600 to 900.

The contradiction between this fact and repeatedly experimentally confirmed regularity of "facilitation" of realization of lyotropic C-state for highly-molecular fractions of polymers can be explained to our mind by the incorrectness of using anomalously low values of sensitivity of shear tension m($\tau=\eta\cdot\gamma m$) as the criterion of the formation of anisotropic phase in solution. The experimental data known to us do not allow to judge with confidence upon the phase state of a deformed system proceeding from the degree of deviation from the Newtonian character of flow: some authors point at a less significant decrease in the viscosity of anisotropic solutions with increasing shear velocity in comparison with isotropic ones [110]; experimental data of other investigators [111] testify to the opposite. Existence of the fluidity limit in the dependence of viscosity on shear tension is a more reliable indication of the anisotropy of a system. Possible interpretations of this phenomenon on molecular level are given elsewhere [102,113].

Thus joint efforts of different investigators allowed to reveal the entity of the rheologic criteria (presence of fluidity limit in the curves of flow as well as the extremal character of the concentration and temperature dependencies of viscosity), the presence of which permits to refer a system to liquid-crystalline ones. Beside absence of characteristic rheologic indicators in some cases gives rise to the doubt as to the reality of C-ordering in polymer solutions (Kulichikhin [102].

Analysis of literature and our own experimental data allows to judge upon the regularities of the influence of solvent nature and composition of the degree of polymerization of cellulose and its derivatives as well as of the character of radicals that are introduced to substitute hydroxyl groups on the polymer conformation in solutions. There is a clear interrelation between the phenomena of salvation and mesomorphism in non aqueous solutions. The results obtained are of essential importance for the perfection of the technologies of production of the cellulose derivatives and products of their processing. Beside one opens the possibilities of the regulation of structure and properties of membranes fibers and liquid-phase thickening agents based on cellulose derivatives for textile industry.

References

1. Krestov G.A., Timofeyeva G.I., Myasoyedova V.V. et.al Dokl. AN SSSR, 1989, V.305, N.1, p.131-134.
2. Gotlib Yu. Ya., Darinskii A.A., Svetlov Yu.E. *Physical Kinetics of Macromolecules*. Leningrad: Khimiya, 1986, p.272.
3. Smirnovd V.N., Iovleva M.M., Kopiev M.A. et. al *Khim. volokna*, 1988, N 3, p.23-24.

4. *Liquid-Crystalline Polymers* Ed. by N.A. Plate. Moscow: Khimiya, 1988, p.416.
5. Khokhlov A.R. *Statistical Physics of Macromolecules*., Moscow: MGU, 1985, 192 p.
6. Kuhn W. *Kolloid Ztschr.*, 1934, Bd. 68, N 1, s. 2-15.
7. Guth E., Mark H. *Monatsh. Chem.*, 1934, Bd. 65, s. 93-105.
8. Marikhin V.A., Myasnikova L.P. *Overmolecular Structure of Polymers*. Leningrad: Khimiya, 1977, 240 p.
9. Eyring H. *Phys. Rev.*, 1932, vol. 39, N 4, p. 746-748.
10. Tsvetkov V.N. *Rigid-Chain Polymer Molecules*. Leningrad: Nauka, 1986, 380 p.
11. Tsvetkov V.N., Shtennikova I.N., Skazka V.S., et al. *J. Polym. Sci. C*, 1968, N 16, p. 3205-3217.
12. Burchard W. *Makromol. Chem.*, 1965, Bd. 88, s. 11-28.
13. Dotu P. *Proc. Nat. Acad. Sci. USA*, 1956, vol. 42, N 11, p. 791-800.
14. Kratky O., Porod G. *Res. Trav. Chim.*, 1949, t. 68, N 12, p. 1106-1122.
15. Porod G. *Monatsh. Chem.*, 1949, Bd. 2, s. 251-258.
16. Rudkovskaya G.D., Shabselie B.M., Baranovskaya I.A., et. al. *Vysokomol. soed.*, 1989, V. 31A, N 1, p.133-139.
17. Tsvetkov V.N., Andreyeva L.N., Lavrenko P.N. et. al. *Vysokomol.soed.*, 1988, V. 30A, N 6, p.1263-1268.
18. Pogodina N.V., Evlampieva N.P., Tsvetkov V.N., et. al. *Vysokomol. soed.*, 1988, V. 30A, N 6, p.1198-1205.
19. Tsvetkov V.N., Andreyeva L.N., Bushin S.V., et. al. *Vysokomol.soed.*, 1988, V. 30A, N 4, p.713-721.
20. Novakovskii V.B., Strelina I.A. *Vysokomol. soed.*, 1988, V. 30A, N 12, p.2467-2472.
21. Yamakava H., Fujii M. *Macromolecules*, 1974, vol. 7, N 1, p. 128-135.
22. Yamakava H., Yoshizaki T. *Macromolecules*, 1980, vol. 13, N 3, p. 633-643.
23. *Cellulose and Its Derivatives* ed. by M. Bickles, L. Segal. Moscow: Mir, 1974, 1010 p.
24. Stockmayer W.H., Fixman M. *J. Polym. Sci. C*, 1963, N 3, p. 197-204.
25. Tsvetkov V.N., Lezov A.V. *Vysokomol. soed.*, 1984, V. 26B, N 7,p. 494-498.
26. Bushin S.V., Tsvetkov V.N., Lysenko E.B. et. al. *Vysokomol.soed.*, 1981, V. 23A, N 11, p.2494-2503.
27. Bianchi E., Ciferri A., Conio J., et al. *Macromolecules*, 1985, vol. 18, p. 646-650.
28. Laivins G.V., Gray D.G. *Macromolecules*, 1985, vol 18, p. 1746-1752.
29. Bohdanecky M., Kovar T. *Macromolecules*, 1983, vol. 6, N 9, p. 1483-1492.
30. Alexeyeva O.A., Timofeyeva G.l., Osipov S.A. et. al. Abstr. All-Union conf. "*Chemistry and Application of Non-Aqueous Solutions*", Ivanovo, 1986, V. 3, p. 437.
31. Archibald W.J. *J. Appl. Phys.*, 1947, vol. 18, N 4, p. 362-367.

32. Popov V.P., Dubrovin V.I., Tarabakin S.V. et. al. *Analytical Ultracentrifugation in Chemistry and Biology*. Dushanbe:1987, p. 87-98.
33. Myasoyedova V.V., Alexeyeva O.V., Krestov G.A. V *All-Union Meeting on Chemistry of Non-Aqueous Solutions of Inorganic and Complex Compounds*. Rostov-na-Donu, 1985, p.226 227.
34. Frenkel S.Ya. *Introduction to Statistical Theory of Polymerization*. Moscow: Nauka, 1965, 267 p.
35. Klenkova N.I., Bludova O.S., Matveeva N.A., et al. *Cell. Chem. and Technol.*, 1982, vol. 16, N 6, p. 615-630.
36. Parkov S.P. *Vysokomol. soed.*, 1982, V. 24B, N 11, p.869-873.
37. Tsvetkov V.N., Andreyeva L.N., Bushin S.V., et. al. *Vysokomol.soed.*, 1988, V. 30A, N.4, p.713-721.
38. Conio J., Bianchi E., Ciferri A. et al. *Macromolecules*, 1983, vol. 16, p. 1264-1270.
39. Kuznetsova G.B., Silinskaya I.G., Kallistov O.V. et. al. *Vysokomol. soed.* 1988, V.30A, N 3, p.586-590.
40. Lyubina S.Ya., Klenin S.I., Strelina I.A. et. al. *Vysokomol.soed.*, 1977, V. 19A, N 2, p.244-249.
41. Swenson H.A., Schmitt C.A., Thompson N.S. *J. Polym. Sci.*, 1965, vol. 11 C, p. 243
42. Satio M. *Polymer J.*, 1983, vol. 15, N 3, p. 213-223.
43. Flory P.J., *Principles of polymer chemistry*, Cornell University Press, 1953.
44. Lyubina S.Ya. *Investigation Methods for Cellulose* Ed. by V.P. Karlivan, Riga: Zinatne, 1981, 264 p.
45. Prozorova G.E., Iovleva M.M., Dibrova A.K. et. al. *Vysokomol.soed.*, 1986, V. 28B, N.6, p.412-415.
46. Pogodina N.V., Melnikov A.B., Mikryukova O.I. et. al. *Vysokomol.soed.*, 1984, V. 26A, N.12, p.2515-2520.
47. Pogodina N.V., Levrenko P.N., Pozhivilko K.S. et. al. *Vysokomol. soed.*, 1982, V.24A, N 2, p.332-338.
48. Bushin S.V., Lysenko E.B., Cherkasov V.A. et. al. *Vysokomol.soed.*, 1983, V. 25A, N.9, p.1899-1905.
49. Pogodina N.V., Bushin S.V., Melnikov A.B. et. al. *Vysokomol.soed.*, 1987, V. 29A, N.2, p.299-305.
50. Pogodina N.V., Pozhivilko K.S., Melnikov A.B. et. al. *Vysokomol.soed.*, 1981, V. 23A, N 11, p.2454
51. Shtennikova I.N. *Investigation Methods for Cellulose* Ed. by V.P. Karlivan. Riga: Zinatne, 1981, 264 p.
52. Pogodina N.V., Pozhivilko K.S., Evlampieva N.P. et.al. *Vysokomol. soed.*, 1981, V. 23A, N 6, p.1252-1260.
53. Shulz G.V., Penzel E. *Makromol. Chem.*, 1968, vol. 112, p. 260-280.
54. Ryumtsev E.I., Aliyev F.M., Vitovskaya M.G. et. al. *Vysokomol.soed.*, 1975, V. 17A, N.12, p.2676-2681.
55. Meyerhoff G., Sutterlin N. *Makromol. Chem.*, 1965, Bd. 87, s. 258-270.
56. Tsvetkov V.N., Lavrenko P.N., Andreeva L.N., et al. *Europ. Polym. J.*, 1984, vol. 20, N 8, p. 823-829.
57. Werbowyj R.S., Gray D.G. Il *Macromolecules*, 1980, vol. 13, p. 69-73.

58. Andreeva L.N., Lavrenko P.N., Urinov E.U. et. al. *Vysokomol.soed.*, 1975, V. 17B, N.4, p.326-330.
59. Shanbhag V.P., Ohman J. *Arkiv Kemi.*, 1968, vol. 29, p. 163-178.
60. Shanbhag V.P. *Arkiv Kemi.*, 1966; Bd. 29, s. 1-22.
61. Burchard W., Husemann E. *Makromol. Chem.*, 1961, vol. 44, p. 358-387.
62. Burchard W. *Br. Polym. J.*, vol. 3, p. 214-220.
63. Musayev Kh.N., Akbarov Kh.I., Nikonovich G.V. et. al. *Vysokomol.soed.*, 1988, V. 30A, N 7, p.1370-1373.
64. Ryskina I.I., Vakulenko N.A. *Vysokomol. soed.*, 1987, V. 29A, N2, p.306-312.
65. Filipp B., Lavrenko P.N., Akatova O.V. et. al. *Vysokomol. soed.*,1987, V. 29A, N 1, p.32
66. Lyulina S.Ya., Klenin S.I., Strelina I.A. et. al. *Vysokomol.soed.*, 1973, V. 15A, N.3, p.691 -698.
67. Korneyeva E.V., Lavrenko P.N., Urinov E.U. et. al. *Vysokomol.soed.*, 1979, V. 21, N.7, p.1547-1553.
68. Zakharova E.N., Kutsenko L.I., Tsvetkov V.N. et. al. *Vestn. LGU, Ser. fiz. i khim.*, 1970, N 16, p.55-63.
69. Tieda D.A., Stratton R.A. *Abstracts. Cellulose, Paper and Textile Chemistry Division* . ACS Meeting, 1977, p. 5-7.
70. Ozolinsh R.E., Kakla M.A. *Khimiya drevesiny*, 1984, N 3,p.68-70.
71. Marrinan H.J., Mann J. *J. Appl. Chem.*, 1954, vol. 4, p. 204-212.
72. Shakhparonov M.I., Zakurdayeva N.P., Pedgorodetskii Yu.K. *Vysokomol. soed.*, 1967, V.19A, N 6, p.1212-1220.
73. Pearson G.P., Moore W.R. *Polymer*, 1960, vol. 1, p. 144-148.
74. Kamide K., Saito M. *Polymer J.*, 1982, vol. 14, p. 517-526.
75. Birshstein T.M., Zhulina E.V. *Polymer*, 1984, vol. 25, N 10, p. 1453-1457.
76. Birshtein T.M., Skvortsov A.M., Sariban A.A. *Structure of Polymer Solutions. Scaling and Computer Simulation*. Preparing of Scient. Cent.Biol. Res., Pushchino, 1981.
77. Baranov V.G., Amribakhshov D.Kh., Brestkin Yu.V. et. al. *Vysokomol. soed.*, 1987, V. 29A, N 6, p.1190-1194.
78. Borisov O.V., Birshtein T.M., Zhulina E.B. *Vysokomol. soed.*, 1987, V.29A, N.7, p.1413-1418.
79. Romamtsova I.I.*Vysokomol. soed.*, 1987, V. 29A, N 1, p.56-61.
80. Pletneva S.G., Marchenko G.N., Pavlov A.S. et. al. *Dokl. AN SSSR*, 1982, V. 264, N.1, p.109-112.
81. Bueche F. *J. Chem. Phys.*, 1954, v. 22, N 9, p. 1570-1576.
82. Vinogradov G.V., Malkin A.Ya. *Rheology of polymers*. Moscow: Khimiya, 1977, 439 p.
83. Philippoff W., Gaskins F.H., Brodhyan J.G. *J. Appl. Phys.*, 1957, vol. 28, N 10, p. 1118-1123.
84. Graessley W.W. *J. Chem. Phys.*, 1965, vol. 43, N 8, p. 2696-2703.
85. Williams M.C. IIA. I. *Ch. E. J.*, 1966, vol. 12, N 6, p. 1064-1070.
86. Middemon S *Flow of Polymers*. Moscow: Mir, 1971, 259 p.
87. Golovko L.I., Rumyantsev L.Yu., Shiliv V.V. et.al. *Vysokomol.soed.*, 1988, V. 30A, N.12, p.2572-2577.

88. Chang Dei Khan. *Rheology in the Processes of Polymer Processing*. Moscow: Khimiya, 1979, 368 p.
89. Vinogradov G.V., Malkin A.Ya., Kulichikhin V.G. *Success in Rheology of Polymers* Ed by V.G.Vinogradova, Moscow: Khimiya, 1970,296 p.
90. Paknshver E.A. *Investigation Methods for Cellulose* Ed. by V.P. Karlivan, Riga: Zinatne, 1981, 264 p.
91. *Superhighmodular Polymers* Ed. by A. Ciferri, I. Ward, Leningrad: Khimiya, 1983, 272 p.
92. Navard P., Haudin J.M. *IUPAC MACRO '83*, Bucharest, 1983.
93. Ciferri A., Valenti B. *Superhighmodular Polymers* Ed. by A. Ciferri, I. Ward, Leningrad: Khimiya, 1983, p.151-168.
94. Bosch T.A., Maissa P., Sixov P. *Il nuovo cimento*, 1984, vol. 3 D, N 1, p. 95-103.
95. Khokhlov A.R. *Liquid-Crystalline Polymers* Ed. by N.A. Plate Moscow: Khimiya, 1988,
96. Kiss G., Porter R.S. *J. Polym. Sci., Polym. Phys. Ed.*, 1980, vol. 18, p. 361-365.
97. Krigbaum W.R. *Polymer liquid crystals.* / Ed. by A. Ciferri, W. R. Krigbaum, N.Y.: Academic Press, 1982, p. 275-308.
98. Keller A. *Superhighmodular Polymers*/ Ed. by A. Ciferri, I. Ward, Leningrad: Khimiya, 1983, p.241-266.
99. Capaccio G., Ward l.M. *Polym. Eng. Sci.*, 1975, vol. 15, p. 219-224.
100. De Gennes P. *J. Chem. Phys.*, 1974, v. 60, N 12, p. 5030-5042.
101. Elyashevich G.K., Frenkel S.Ya. *Orientation Phenomena in Polymer Solutions and Melts*. Moscow: Khimiya, 1980, p.9-90.
102. Kulichikhin V.G. *Liquid-Crystalline Polymers*/ Ed. by N.A. Plate, Moscow: Khimiya, 1988, p.331-371.
103. Tsvetkov V.N. *Vysokomol. soed.*, 1979, V. 21A, N 11,p.2606-2623.
104. Gilbert R.D., Patton P.A. *Prog. Polym. Sci.*, 1983, vol. 9, p. 115-131.
105. Dayan S., Gilli M., Sixov P. *J. Appl. Polym. Sci.*, 1983, vol. 28, N 4, p. 1527-1534.
106. Conio J., Bruzone P., Cifferi A. *Polym. J.*, 1987, vol. 19, N 6, p. 757-768.
107. Suto S. *J. Appl. Polym. Sci.*, 1987, vol. 34, p. 1773-1776.
108. Krestov G.A., Myasoyedova V.V., Alexeyeva O.V. et. al. *Dokl. AN SSSR*, 1987, V. 293, N 1, p.174-176.
109. Patel D.L., Gilbert R.D. *J. Polym. Sci., Polym. Phys. Ed.*, 1981, vol. 19, p. 1449-1460.
110. Kulichikhin V.G., Petrova L.V., Khanchich O.A. et. al. *Khim. Volokna*, 1985, N 2, p.42 p.135-139.
111. Belov S.Yu., Myasoedova V.V., Krestov G.A. *Zhurn. prikl. Khimii*, 1989, N 1,
112. Papkov S.P., Kulichikhin V.G. *Liquid-Crystalline State of Polymers*. Moscow: Khimiya, 1977, 240 p.
113. Asada T., Onogi S. *Polym. Eng. Rev.*, 1983, v. 3, N 2-4, p. 323-353.

COMPUTER SIMULATION OF ASSEMBLIES OF RIGID ELASTIC ELLIPTIC PARTICLES

A.A. Berlin*[1], L. Rothenburg, and R.J. Bathurst*****

*N.N.Semenov Institute of Chemical Physics, Russian Academy of Sciences, Kosygina St. 4, Moscow, 117977 Russia

**University of Waterloo, Ontario, Canada, N2L3G1

***Royal Military College of Canada, Kingston, Ontario, Canada, K7K5L0

Abstract - Results of computer simulation of static mechanical behavior of a body consisting of ~1000 hard elastic particles have been invoked to discuss some problems of mechanics of disordered (amorphous) and ordered (crystal) bodies: the glass-liquid translation, irreversible deformation (plasticity in a solid and flow in a liquid states), intermediate (like liquid crystal) state.

It has been shown that the existence of two states, solid and liquid, the condition of transition between them, and the fundamental mechanical properties of a solid body, viz., plasticity, strain softening, localization of deformation within the shear bands, are controlled by the number of physical contacts between the particles, spatial distribution of contacts, and their disintegration under shear deformation.

Systems of rigid particles are in solid or liquid states in dependence on the number of interparticle contacts. Solid (glass or crystal) and liquid states were determined from the ability of a system to resist the shape change under external force. As a criterion of the liquid to glass transition the equality of the number (translational and rotational) degrees of freedom, f, and the number of constraints of these motions, C, was used (f=C). System is solid if f< C, and is liquid if f>C. In systems of rigid particles constraints are due to mechanical contacts (C_1) and/or chemical bonds between the particles (C_2). Glasses were classified as mechanical (granular systems and

---

[1] Permanent address: Al.Al.Berlin, Division of Polymers and Composite Materials, Institute of Chemical Physics of the Russian Academy of Science, Kosygin Str.,4, Moscow, 117977, Russia.

metallic glasses) if $C_1>>C_2$, chemical (non-organic glasses) if $C_1<C_2$, and combined (polymer glasses) if $C_1\approx C_2$. Some results of computer imitation confirming the transition criterion are presented.

Irreversible deformation of the particles assemblies in liquid state unlike in solid one showed the following features: extremely low yield stress, the number of interparticle contacts and the volume of the system do not change during flow, distribution or local strains is random. Shear strain of the liquid assemblies is governed by particles rotation and consists in the changes of particles orientation while shear strain of glassy solid is due to the disintegration of interparticles contacts. A relationship between distribution function of particles orientation and shear strain of the body is found.

Crystals of rigid elliptic particles are anisotropic and demonstrate solid-like, liquid-like or intermediate behavior in different directions depending on its arrangement and ellipse eccentricity.

Validity of the geometrical concepts for developing the theory of a condensed state is based on the assumption that any interaction potential can be decomposed into two components: hard repulsion and soft attraction, which are responsible for various properties of the materials.

1. INTRODUCTION

All bodies consist of particles, which interact with each other through various forces described by different interaction potentials. At high densities in the condensed state the hard repulsive forces play a key role and control some important properties of the condensed media, e.g., crystalline structure with its defects, etc.

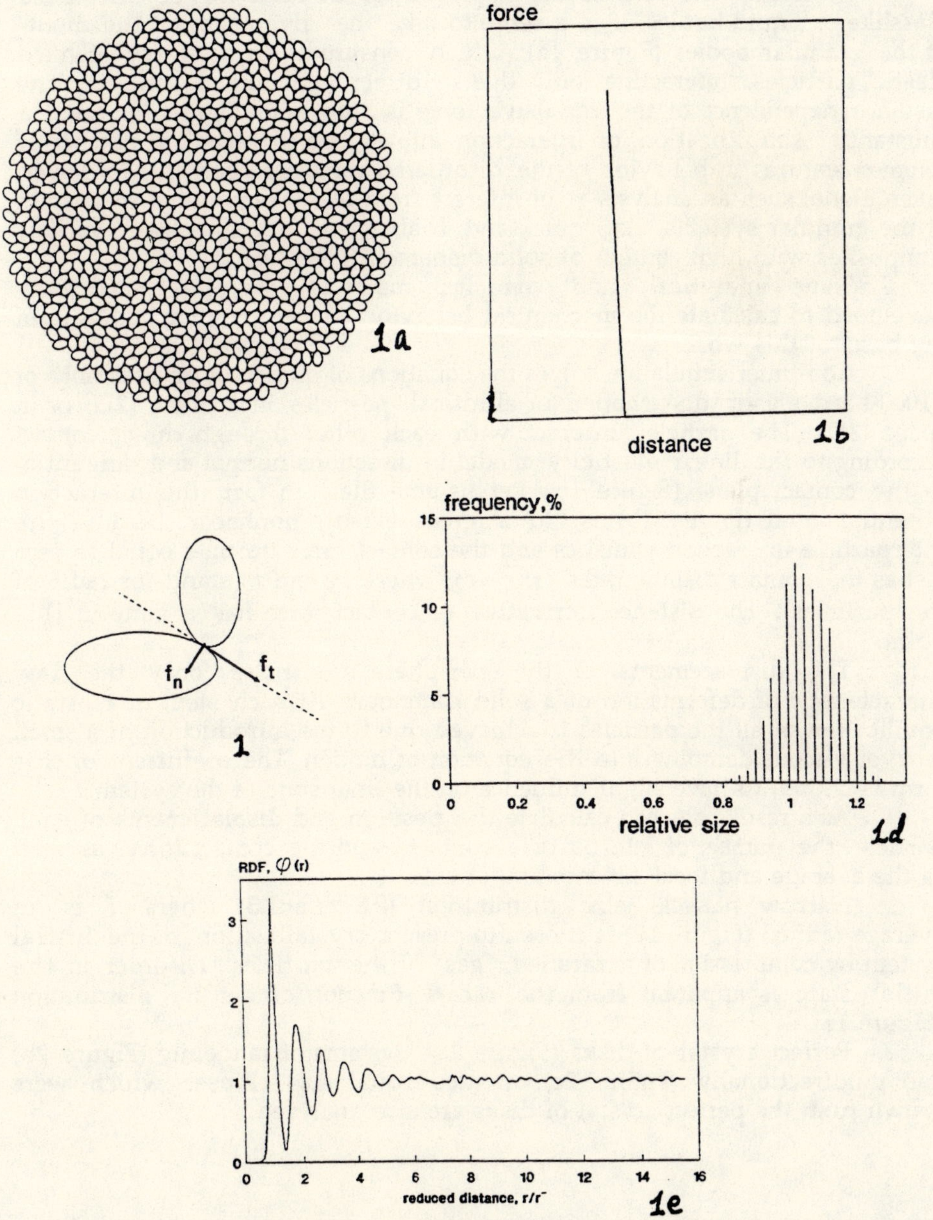

Figure 1. Representation of a ellipses assembly analyzed (a); dependence of contact forces on distance between the centers of particles (b); scheme of normal, f_n, and tangential, f_t, contact forces (c); function of particle size distribution (d); and normalized RDF (e).

To clarify the role of the above forces in behavior of disordered glasslike or liquid bodies, it is a sense to take the advantages of the model of the granular bodies (Figure 1a), which constitutes an assembly of hard elastic particles, interacting only due to direct geometrical contacts. The distance dependence of the repulsive force is shown in Figure lb. Such a substantial simplification of interaction allows one to reveal a number of common features in behavior of the disordered condensed bodies. More over, taken alone, such an analysis is of interest from the viewpoint of mechanics of the granular systems, viz., quicksand, friable soil, dry substances, various composites with high content of solid dispersed fillers, etc.

Some analytical and computer methods (see [1-11]) already developed to calculate the mechanical behavior of the granular bodies form the basis of this work.

Computer simulation solves the equations of motion of an assembly of ~1000 hard elastic disk-shaped or elliptical particles in a plane (2D) or in space (3D). The particles interact with each other through direct contact according to the linear elasticity model in directions normal and tangential to the contact plane (Figure 1c). Let us note that, in fact, the interaction potential used (Figure 1b) is, to a great extent, nonlinear, because the interparticle interaction vanishes and the contact force becomes equal to zero just as the contact disintegrates ($r>r_1+r_2$, where r_1 and r_2 stand for radii of the particles). The distance derivative of contact force has a jump in this point.

The displacements of the peripheral particles obey the law characteristic of deformation of a solid continuum. At each step, quasi-static equilibrium of all the particles is achieved due to the introduction of a small term of viscous damping into the equation of motion. The coefficient of this term is chosen to have slight influence on the final state of the system.

As a result, one can calculate the position and displacements of each particle, the number of interparticle contacts, and the contact forces as well as the average and local stresses and strains.

Narrow particle size distribution ($0.8<r/\bar{r}<1.3$, where $\bar{r}$ is an average radius) (Figure 1d) is chosen to present crystallization of the initial system by compression of a rarefied "gas" of the particles. Disorder in the initial state is apparent from the radial function of density distribution (Figure 1e).

Perfect crystal of disks (Figure 2a), systems of randomly (Figure 2b) and unidirectionally (Figure 2c) oriented equal size ellipses which were grown from the perfect crystal of disks are also analyzed.

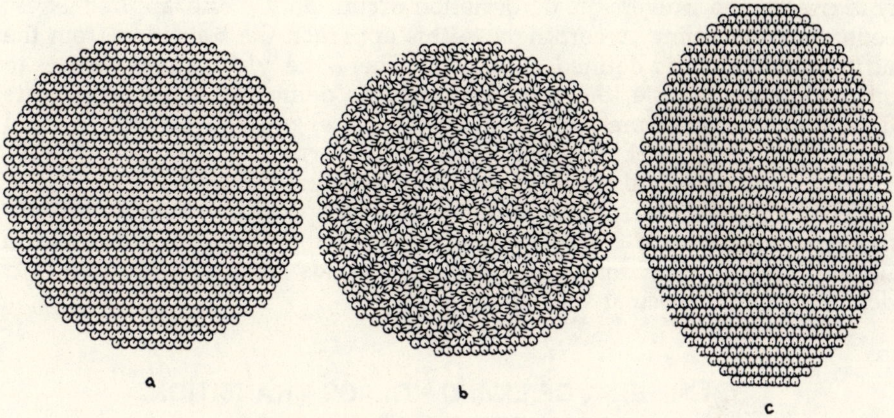

Figure 2. Representation of perfect crystal of disks (a); assemblies of randomly (b) and unidirectionally oriented equal ellipses (c).

Average density of the bodies studied is rather high and equal to 0.8-0.9.

2. LIQUID AND SOLID STATES.

Thermodynamic definition of liquid and glassy states is usually involves the problem of transition from ergodic (liquid) to non-ergodic (glassy) system. Similar problem appears in granular materials which are studied by themselves and, in addition, are a convenient model of molecular body. Granular materials are always non-ergodic. In this case the thermodynamic approach is applied only with some doubtful suppositions such as permanent exchange of neighboring particles [12].

The definitions of liquid and glass states based on the difference of their relaxation properties have the following disadvantages. The spectrum of relaxation times is wide both in liquid and solid states. Some motion modes may be frozen in liquid state, the other may be free in solid state. Moreover, the transition between these states depends on the test rate and time of experiment.

That is why the liquid and glass states are defined here on the basis of their mechanical behavior.

The difference between liquid and solid states consists in the ability of the solid body to resist the shape change under external forces. The body is liquid if it deforms irreversibly under negligibly low applied stress. The body is rigid if the irreversible deformation occurs only if the applied stress exceeds the yield point. Accordingly to this approach the transition from the liquid to glassy state is defined as a point where the yield stress reduces to zero [13-16]. For example, the transition may be defined if some parameters such as temperature or pressure are varied. For a granular system of rigid particles this parameter is the number of interparticle contacts [13-17]. Systems of unbinded rigid particles were considered in [13-16].

In this paper the approach developed is applied to various systems of rigid particles which, in addition to mechanical (repulsive) contacts, are connected by rigid directional like chemical bonds. Such system may be a model for various molecular substances.

3. Criteria of Liquid - Glass Transition, Classification of Glasses

The model is based on the suggestion that the system is solid if the particles lose translational and rotational mobility [13-15]. The parameter which determines the state of a substance (liquid or solid) is the difference between the number of freedom degrees (translational and rotational), f, and the constraints of these motions, C [13-15].

Computer imitation of mechanical behavior of rigid elastic particles with narrow size distributions showed that the liquid to glass transition is observed when the number of freedom degrees is equal to the number of constraints [13,14]:

$$f = C \tag{1}$$

This conclusion is the foundation for further analysis. The system is solid if $f<C$ and liquid if $f>$ C. An example of mechanical stability of a rigid beam illustrates this criterion (Figure 3). Some data of computer imitation of mechanical behavior of hard elastic particles confirming the above assumption (Equation 1) are presented below.

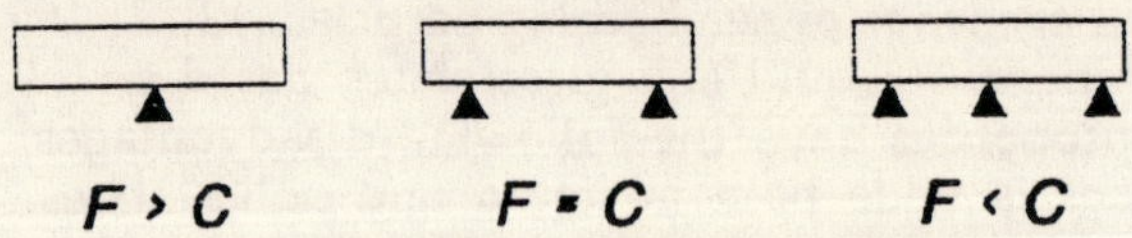

Figure 3. Illustration of liquid - solid transition for mechanical system like beam on fulcrums.

The second assumption is that in the system of solid particles with chemical bonds between particles (Figure 4) the additional constraints are due to rigid chemical bonds. The number of constraints is equal to the sum of mechanical (C_1) and chemical (C_2) constraints:

$$C=C_1+C_2 \tag{2}$$

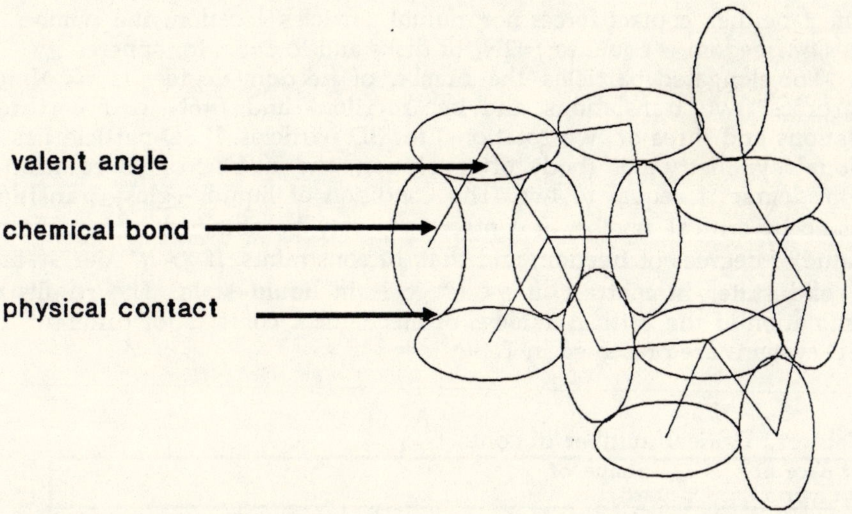

Figure 4. System of interconnected particles. I-mechanical contact, II-chemical bond, III-rigid angle.

Hence all glasses may be divided into three groups.

1. Glasses with random mechanical contacts ($C_1>>C_2$). Such systems are granular bodies and, possibly, metallic glasses.

2. Glasses with chemical bonds ($C_1<<C_2$). The example is non-organic glass.

3. Glasses with both mechanical and chemical constraints ($C_1 \approx C_2$). Examples are polymer and low molecular organic and non organic glasses. The first type of glasses may be called "mechanical", the second type "chemical" and the third "combined glasses."

Let us apply the criterion of liquid to glass transition (Equation 1) to every of these types of systems.

1. Mechanical glasses. In this case particle motion is restricted only by mechanical contacts. If particles are smooth spheres or disks, the repulsive forces are central and the particle displacement perpendicular to the contact plane is forbidden. The number of constraints, C_1, in the system of particles is equal to the number of contacts $C_1=\gamma N/2$, where γ is the average number of contacts.

If high friction* between particles occurs, their mutual shift is not possible and the number of constraints is $C_1 = \gamma N$ for two-dimensional (2D) and $C_1 = 3\gamma N/2$ for three-dimensional (3D) systems.

As the rotation of smooth disks (2D) and smooth spheres (3D) does not affect neither contact forces nor mutual particles location, the number of degrees of freedom is equal to $f=2N$ for disks and to $f=3N$ for spheres.

For elongated particles the number of freedom degrees is $f=3N$ for 2D-particles (two translations and one rotation) and $f=6N$ or 5N (three translations and three or two rotations) for 3D particles. If 3D-particle has a rotational symmetry axis (body of revolution) the number of its rotational freedom degrees is equal to two. The condition of liquid - glass transition expressed by critical number of contacts, γ^*, may be obtained by equalizing the value of degrees of freedom and that of constraints. If $\gamma > \gamma^*$ the system is in solid state. In contrast, if $\gamma < \gamma^*$ it is in liquid state. The results of determination of the critical number of mechanical contacts for different 2D and 3D systems are presented in Table 1.

Table 1. Critical number of contacts, γ^*.

Space of dimension	Shape of particles	C	f	γ^*
2	Smooth disk	$\gamma N/2$	2N	4
2	Rough disk	γN	3N	3
2	Smooth ellipse[1]	$\gamma N/2$	3N	6
2	Rough ellipse	γN	3N	3
2	Smooth polygon[2]	gN	3N	3
2	Rough polygon[2]	$3\gamma N/2$	3N	2
2	Smooth sphere	$\gamma N/2$	3N	6
3	Rough sphere	$3\gamma N/2$	6N	4
3	Smooth ellipsoid[1]	$\gamma N/2$	6N(5N)[3]	12(10)[3]
3	Rough ellipsoid[1]	$3\gamma N/2$	6N	4
3	Smooth polyhedron[4]	$2\gamma N$	6N	3
3	Rough polyhedron	$3\gamma N$	6N	2

1 - or a body (figure) of arbitrary shape with point contacts;
2 - linear contact only;
3 - for body of revolution;
4 - plane contacts only.

It is worth mentioning that contacts may be point, linear and plane. For such particles as disks, spheres and ellipsoids contacts are point. For particles with plane boundaries (polygons, polyhedra) contacts may be

* The terms "high friction" and "rough body" here and below mean the lack of particles sliding in contact and existence of both normal as well as tangential elastic contact forces.

linear and plane as well. Plane and linear contacts allow slipping along the contact, while the mutual rotation of particles is forbidden. Hence, the number of constraints, C, in this case is higher in comparison with point contacts.

The critical number of contacts is equal to $\gamma^*=3$ three for smooth 2D polygons with linear contacts and to $\gamma^*=6$ for smooth ellipses. Similar results can be obtained for particles of more complex shape.

Data of computer simulation represented below confirm the accuracy of the above equations for the critical contact numbers (Table 1) and the transition criterion. The method and results of the computations are described in [3,5-11].

Figure 5 shows stress-strain curves for biaxial compression ($\sigma_{11}>\sigma_{22}$ = Const) of 2D random system of elliptic particles with eccentricity e of 0.2. Curves 1,2 and 3 correspond to systems of smooth (no friction) particles with the critical number of constraints $\gamma^*=6$, rough (high friction coefficient) ellipses with $\gamma^*=3$ and smooth ellipses (frozen rotation) with $\gamma^*= 4$. At the absence of shear load initial number of contacts is equal to $\gamma^o=5.8$ for every system.

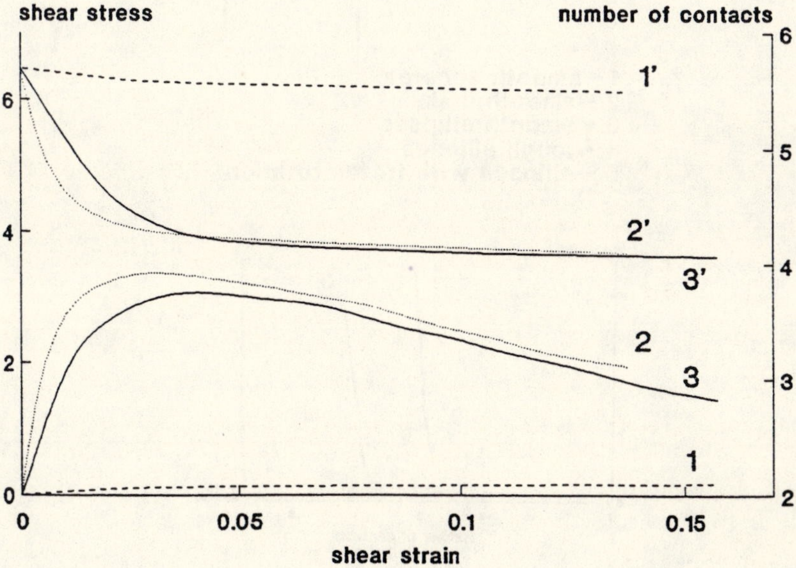

Figure 5. Stress-strain curves (1,2 and 3) and dependence of number of contacts on shear strain (1',2' and 3') for biaxial compression of plane (2D) system of disordered elliptic particles (eccentricity e=0.2). 1,1'- smooth particles (no friction); 2,2'- rough ellipses (strong friction); 3,3'- smooth ellipses with frozen rotation.

This value is close to the theoretical one for the infinite system of elliptic particles ($\gamma^*=6$, see Table 1). Some discrepancy between the real number of contacts $\gamma°$ and the theoretical value is due to the presence of boundary particles with the less number of contacts [16].

In agreement with the above criterion ($\gamma°<\gamma^*$) the yield stress for the system I is negligibly low, and the system deforms as a liquid. In contrast, the systems 2 and 3 deform as solids with high yield stress because the number of constraints exceeds the critical value, $\gamma°>\gamma^*$.

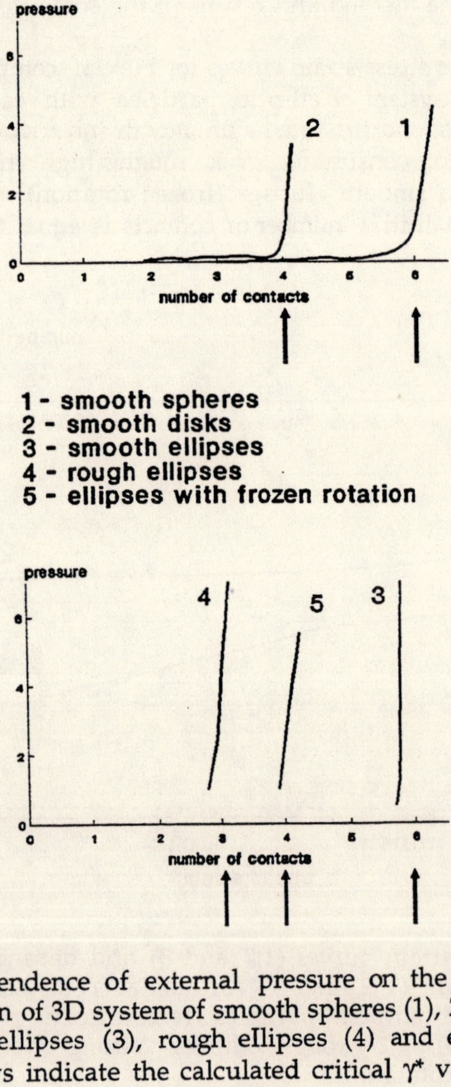

Figure 6. Dependence of external pressure on the number of contacts during consolidation of 3D system of smooth spheres (1), 2D system of smooth disks (2), smooth ellipses (3), rough ellipses (4) and ellipses with frozen rotation (5). Arrows indicate the calculated critical γ^* values (Table 1).

Figure 6 shows the results for hydrostatic compression of initially loose system of particles. Initially, the number of contacts grows at practically zero external pressure. When the number of contacts reaches the critical value, γ^*, a sharp increase in the external pressure is observed. Hence, in this point the transition from liquid to solid is observed. Figure 6a shows the results for smooth disks (2D system, $\gamma^*=4$) and smooth spheres (3D, $\gamma^*=6$). Figure 4b shows the results for 2D systems of smooth ellipses ($\gamma^*=6$), ellipses with strong friction ($\gamma^*=3$) and smooth ellipses with frozen rotation ($\gamma^*=4$). The results of computer calculations are in good agreement with proposed analytical predictions (Table 1).

Let us note two important features of the liquid-glass transition.

a. Density of a system at the transition point is not an intrinsic and characteristic parameter; it depends on the shape of the particle and on nature of the contact forces. For example, at the transition point the density of system of the smooth disks in a plane is equal to ≈ 0.8, and of the smooth ellipses is equal to ≈ 0.9. There are also some other examples that demonstrate the inadequacy of the simple models of free volume or critical pore concentration.

b. The above data show that the change in the particle shape symmetry (disks or ellipses) or in the symmetry of the contact forces (normally or arbitrarily directed) leads to a marked and sharp change in the properties of the system at the transition point. It allows one to draw an analogy between the nonrelaxational liquid-glass transition and the thermodynamic second-order phase transition.

2. Chemical glasses. Let us consider systems with rigid directional chemical bonds as additional constraints of other type. Similar approach was described in [18-20], where a relationship between the network structure of non-organic materials (halkogenids) and their ability to crystallization and vitrification was discussed. The problem of competition between crystallization and vitrification will not be concerned in this paper.

Let us consider the constraints introduced by rigid directional chemical bonds. If the particle has M bonds with its neighbors, the number of constraints due to fixed distance between particles is equal to the number of chemical bonds $C_1=MN/2$. The number of constraints related with fixed angles between particles C_2 is given by:

$C_2 = (M-1)N$ for two-dimensional systems; and

$C_2 = (2M-3)N$ (at $M \geq 2$) for three-dimensional systems.

The criterion of liquid to glass transition is given by equation: $f = C_1 + C_2$.

For an example a mixture of two atoms with different chemical valences (M_1 and M_2) is considered. This system corresponds to non-organic glasses such as As_xSe_{1-x}, Ge_ySe_{1-y} etc.

The liquid to glass transition in these systems may be obtained by variation of the mixture composition (x,y) or the fraction of the bond saturation α (α gradually increases with a decrease in temperature).

The glass transition criterion is given by:

$$f = M^*N/2 + (2M^*-3)N = (2,5M^*-3)N$$

In 3D system of smooth spheres $f = 3N$ and hence the critical average valence is $M^* = 2.4$.

For alloy of two atoms with saturated chemical bonds is obtained: $M = M_1X + M_2(1-x)$, and the critical content of the components is given by:

$$X^* = (2,4 - M_2)/(M_1 - M_2).$$

The mixture is in glass state if:

$$x > x^* \text{ for } M_1 > M_2$$

For example, for alloy As_xSe_{1-x} with valences of $M_1=3$ and $M_2=2$ the critical $x^* = 0.4$. Similarly, for alloy Ge_ySe_{1-y} ($M_1=4$ and $M_2=2$) the critical content y^* is equal to 0.2. This conclusion agrees with the results reported in [18-20].

If the chemical bonds exist only between different types of atoms, their number is equal to the minimum of two values, $2M_1Nx$ and $2M_2N(1-x)$. Then the glass transition criterion is:

$$\{2M_1x^*, 2M_2(1-x^*)\}\min=2,4.$$

In this case the composition has two critical contents, x_{min} and x_{max}. The system is in glassy state if $x^*_{min} < x < x^*_{max}$.

For a composition with the valences of atoms $M_1=4$ and $M_2=2$ the critical contents are equal to: $x^*_{min} = 0,3$ and $x^*_{max} = 0,4$. Thus, the composition is in glassy state if $0.3 < x < 0.4$.

If $M_1=3$ and $M_2=2$ then $x^*_{min} = x^*_{max} = 0.4$. Hence, the material is in glassy state only if $x=0.4$.

The critical coefficient of bonds saturation α^* for substance with single type of atoms is equal to: $\alpha^*=2,4/M$, where M is the average number of chemical bonds per particle. For alloy of two types of atoms M is equal to average valence.

It should be mentioned that the latter results are obtained under the assumption that the number of valences (M) is higher or equal to two. For randomly interconnected particles and incomplete bonds saturation for some particles M can be less than two.

3. Combined glasses. Polymer glasses are typical example of such systems. If the chemical valence is M and the number of mechanical contacts per particle is γ, the total number of constraints is equal to: $C= \gamma N/2 + (2,5M-3)N$, and the critical number of physical contacts in liquid - glass transition point for smooth 3D spheres is equal to: $\gamma^* = 12 - 5M$ $(M\geq 2)$.

Let us consider polymer network with valences of atoms equal to two and three (M_1=3 and M_2=2). If the branching degree is the critical number of mechanical contacts is: $\gamma^* = 2 - 5\beta$.

If the branching degree exceeds 0.4, polymer is a glass even at zero number of mechanical contacts. In fact, in highly cross linked polymer networks α–peak (the glass transition) is not observed up to temperatures of chemical degradation [21].

The present results are obtained under the restriction of the motion by neighboring particles. Long-distance interactions contribute additional constraints. It leads to freezing of the molecular rotations around the chemical bonds. This case was analyzed in [14].

Let us consider the effect of the chain rigidity on the glass transition. With this aim, the system of N 3D fragments consisting of *n* smooth spheres with rigid bonds and different angle rigidity is analyzed.

Suggesting that atoms are spherical the number of freedom degrees for the chain is equal to $f = 3Nn$.

The number of mechanical contacts is $C_1 = \gamma N n/2$, where γ is the average number of contacts per a sphere.

The number of chemical constraints is equal to $C_2 = (n-1)N$.

The number of angles constraints is equal to $C_3 = (n-2)N$ (at $n \geq 2$).

The number of rotation constraints is $C_4 = (n-3)N$ (at $n \geq 3$)

Let us consider three types of polymer chain with different rigidities.

1. Flexible chain without angle constraints (free joined chain). Transition criterion is expressed by the equation $f = C_1 + C_2$, which gives the critical number of mechanical contacts as $\gamma^* = 4 + 2/n$. 2. Semirigid chain with fixed valent angles and free rotation around chemical bonds. In this case transition criterion is given by $f = C_1 + C_2 + C_3$, and the critical number of contacts is $\gamma^* = 2 + 6/n$ ($n \geq 3$).

3. Rigid chain with forbidden molecular rotation. Transition criterion is $f = C_1 + C_2 + C_3 + C_4$, and the critical number is $\gamma^* = 12/n$ (at $n \geq 3$).

Thus, the critical number of mechanical contacts at glass transition depends on molecular chain rigidity. For a long rigid chain the physical contacts are not necessary. Since the rigidity of polymer chain increases with temperature decrease, glass transition is observed if a polymer is cooled. The glass transition temperature depends on the activation energy of internal rotation and the energy difference between the energy states.

In conclusion of this section, let us discuss simple, but multicomponent, granular systems. So far, only the systems with a relatively narrow particle size distribution and, what is of particular importance, the systems with an initially (prior to deformation of the system) narrow contacts number distribution of the particles were under the analysis. However, one can regard a bicomponent or even more complex granular system. In this case, an analysis of the average number of interparticle contacts, γ, seems to be unreasonable, and one should consider two (or more) separate subsystems.

Let us now discuss, as an example, a mixture consisting of a great number of bulky spherical particles (or disks) with the characteristic $\gamma > \gamma^*$ and a few small ($r_2/r_1 < 0.1$) spherical particles (or disks) with the characteristic $\gamma < \gamma^*$, which are located between the bulky particles [14]. Such a system is described by a broad and, in this case, bimodal function of the contacts number distribution, and can be defined as a "two phase" system (the latter term has nothing in common with an ordinary two-phase system, which is known to consist of the two individual phases separated by an interface), where the bulky particles are in a solid state and the small particles are in a liquid state. This system experiences two different liquid-glass transitions associated with the above subsystems.

The problems concerning behavior of such systems, quantitative criterion of the allowed separation into the subsystems, analysis of their mutual influence on the properties of the subsystems, including the liquid-glass transition, analysis of the possible mutual dissolution and formation of mixed single component systems are of fundamental interest and require a greater understanding from our point of view.

Let us underline fascinating analogy inherent to molecular systems. Polymers are known to experience a series of so called "relaxation transitions" (α, β, γ, etc.). The high temperature α-relaxation is assigned to the glass transition of a polymer; other transitions are associated with different molecular groups (side groups, etc.). It is worth noting that the approach advanced may be helpful in development of a new theory that describes the above transitions.

Now two remarks concerning glass properties. Low temperature anomalies of heat capacity and acoustic spectrum are clear examples of the difference between glass and crystal. These effects are usually described by the model of two energy states [22-24]. Accordingly to the present model any glass includes some mechanically unstable particles with lower number of contacts. For example in 2D system of disks with average number of contacts $\gamma=4$ approximately two percents of particles has only one or no contacts at all. These mechanically unstable particles are believed to be responsible for low temperature anomalies in glass.

Relaxation properties also show some peculiarities near glass transition temperature. Translational and rotational mobilities of the particles decrease with temperature. As a result the number of unfrozen freedom degrees also decreases. The similar problem was analyzed in [25]. As a result, the rate of relaxation reduces and relaxation times increase up to infinity. Therefore, the glass structure depends on a time of experiment and a test rate. Moreover, near the glass transition point the relaxation rate is usually described by logarithmic law rather than exponential law due to significant reduction of relaxation rate near the glass transition point. These effects are observed both in experiments with real polymers and in computer ones with rigid particles systems. This problem will be thoroughly discussed in following publications.

In conclusion it should be emphasized the phenomena described is a geometrical one, and is due to the existence of the hard repulsive branch of interparticle potential.

4. Mechanism of Irreversible Deformation (Plasticity) in Solid (Glassy) State.

The ability to undergo plastic deformations is one of the most typical and important properties intrinsic to almost all solid bodies, both crystalline and amorphous. It is the plastic deformations that control a series of fracture properties, such as crack resistance (toughness), impact and fatigue strength, etc.

Dislocations growth and shift is known as the only microscopic mechanism responsible for the plastic deformation in the crystalline media. Similar macroscopic plastic behavior of crystalline and amorphous materials had provoked some attempt to extend the like-dislocation models to amorphous bodies [26, 27]. Such extensions, however, lead to serious contradictions.

On the basis of computer simulation [13,14], we managed to identify an alternative mechanism for the plastic deformation in a solid disordered granular system. The main point of our approach involves the change in the average number of interparticle contacts, γ, (contacts, but not neighboring particles!) during deformation in the presence of the shear component, as well as the evolution of the spatial distribution of the particles with a different (less than γ^*) number of contacts. Further analysis must confirm or refute an extension of the concepts proposed to the mechanism of plastic deformation in the wide class of disordered systems. It is most likely in metallic glasses [28,29].

Let us more closely consider the deformation of a body consisting of a granular material at $\gamma > \gamma^*$, i.e., in a solid glass like state. At small deformation an elastic behavior is observed; further deformation produces a plastic state. The plot of the shear component of the mean stress versus the shear strain (σ - ε diagram) is typical for most of the solid bodies, viz., an asymptotic curve or a curve passing a maximum (Figure 5). To explain the strain softening of the material, i.e., the drop in the instant modulus, related to the plastic deformation, consider first a macroscopic manifestation of plasticity inherent to the most of the materials, namely, inhomogeneous deformation localized in the shear bands. As shown in [30], the initiation of the shear bands is attributed to the maximum on the α–ε diagram. Near the maximum, the homogeneous deformation of the sample (under uniform boundary conditions, for example, under uniaxial compression) becomes unstable and localizes within the shear bands oriented approximately along the principal tangential stresses.

Therefore, whatever the microscopic mechanism of plasticity inhomogeneous deformation localized within the shear bands should also occur for the granular systems discussed. Nucleation and growth of inhomogeneous deformation as shear bands are really observed in the region corresponding to the plastic deformation (Figure 7). If the shear band propagates throughout the sample or meets another one, its internal deformation becomes arrested and a new shear band nucleates (Figure 7). Such behavior was observed in the following loading conditions: displacements of the boundary particles were set according with macrodeformation of the whole body.

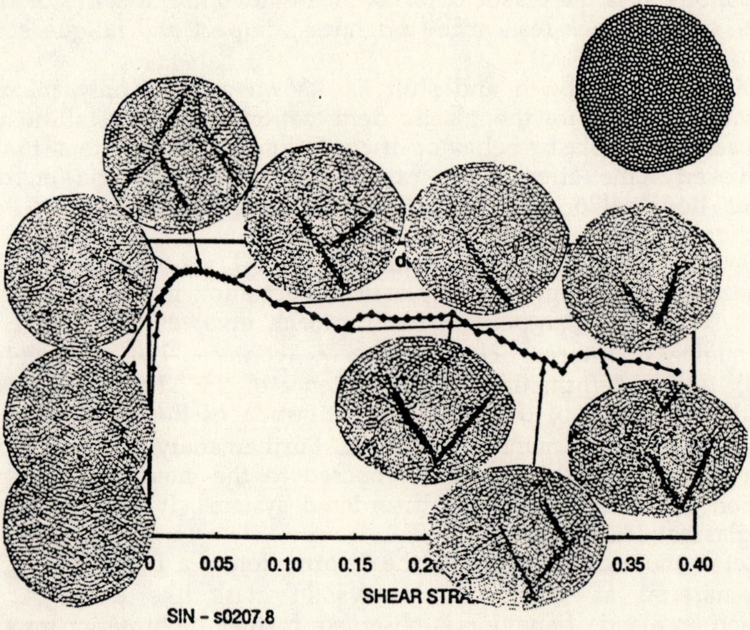

Figure 7. Distribution of local instant strains at various stages of biaxial compression. Direction and length of each dash correspond to direction and displacement of the two neighboring particles at each deformation step. Arrows indicate those points in the compression diagram that correspond to the relevant patterns of strain distribution.

It is important to note that, during the plastic deformation up to ε = 40% (the right limit of our experiments), exchange of the neighbors doesn't occur. Moreover, there is no exchange even at the early stages of the plastic deformation (ε = 1-3%). Over the entire deformation range slight displacements of the particles occurs only, i.e., a slight change of the first coordination sphere shape takes place. Therefore, the mechanism of jumps of the particles over an energy barrier from one equilibrium position to the next (the Eyring model, the free volume model, etc.) seems to be out of question. Another noteworthy side is that, over the entire plastic deformation range, there is no change even in the number of neighbors that occupy the first coordination sphere surrounding each particle.

Similar to the characteristic features of liquid-glass transition the strain softening of a solid granular body is found to be accompanied by the change of interparticle contacts number the preceding section. Under compression or shear stress, the system (a solid glasslike specimen with $\gamma > \gamma^*$) experiences a sharp decrease of the contacts number (Figures 5, 10). Under uniaxial compression, the volume of the system initially decreases slightly and then increases [13,14] and Figure 21 in next chapter. The average number of contacts tends to γ^*. Let us note that, during deformation of liquids ($\gamma < \gamma^*$), no change volume and contacts number occurs (Figure 5, curve 1').

It should be mentioned that the onset of the plastic deformation in a solid body, the position of the maximum in the α–ε curve, and the nucleation of the shear bands are observed when the average number of contacts considerably exceeds γ^* still. The spatial distribution of the particles with the small number of contacts ($\gamma \geq \gamma^*$) is found to be of particular importance. At the first stages of elastic deformation particles with $\gamma \leq \gamma^*$ randomly and homogeneously appear in the sample volume. These particles are then grouped together, creating linear clusters disposed according to the directions of the principal shear stress (Figure 8b). Just along these structures, the principal shear deformations are localized (Figure 8a). It is interesting that the intact contacts within the shear band are preferentially oriented across the shear band axis (Figure 8c) in such a manner that the contact forces do not restrict the shear of one part of the sample along the band with respect to the other.

So, plastic deformation in a solid disordered (glass) body as well as the liquid-glass transition is governed by interparticle contacts number and its critical value, γ^*. Undoubtedly, this phenomenon is in a close connection with a strong repulsive potential assumption.

In fact, any interaction potential in a molecular system can be divided into two branches: the left-hand "hard" repulsive branch and the right-hand "soft" attractive branch. The latter can be invoked to explain such phenomena as the gas-liquid transition, existence of a liquid state and critical points, etc. The "hard" branch, accordingly to our opinion, is a main one for explanation of liquid-glass transition, mechanical behavior of glasses, etc. This part of the interaction potential that controls the structure, defects, and mechanical properties of crystals.

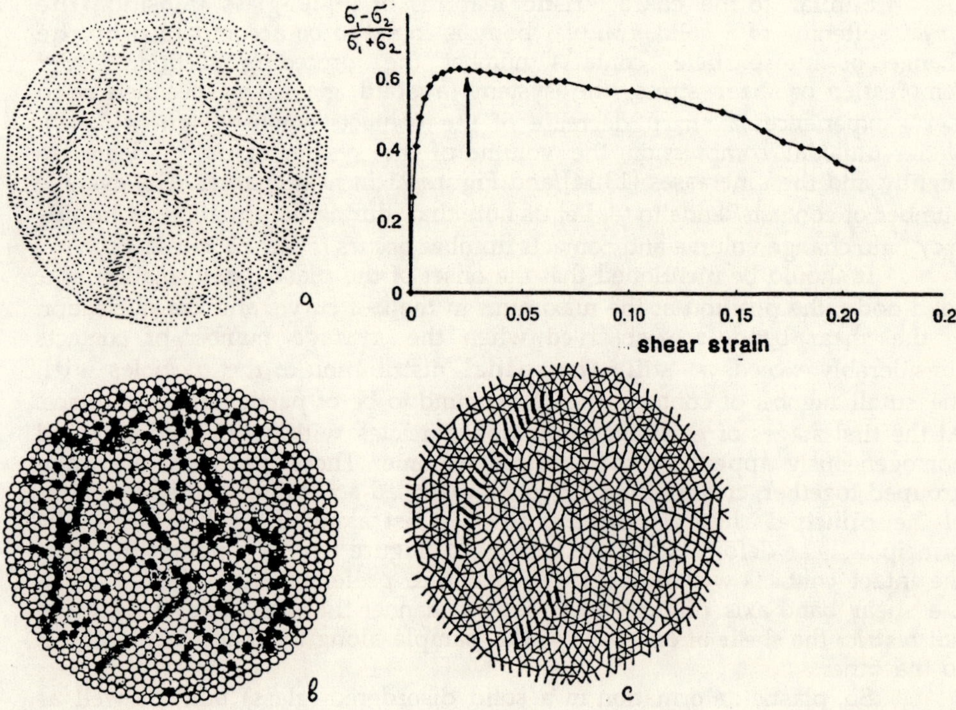

Figure 8. Spatial distributions of local strains (a), particles (dark regions) with the number of contacts $\gamma < \gamma^{*}=3$ (b), and directions of all the interparticle contacts (c). Arrow shows the region corresponding to the above distributions, a, b, and c.

The most complicated problem in the development of the geometrical theory of glasses and liquids within the framework of the approach advanced is a definition of the direct contact, determination of their number a given particle in dependence on external conditions (temperature, pressure, etc.).

Described mechanism of amorphous solid (glassy) plastic deformation - disintegration of interparticle contacts, - is also observed in perfect (without dislocation and other defects) crystal of smooth disks (Figure 9) and ellipses under compression along minor axes of ellipses [16] (Figure 9).

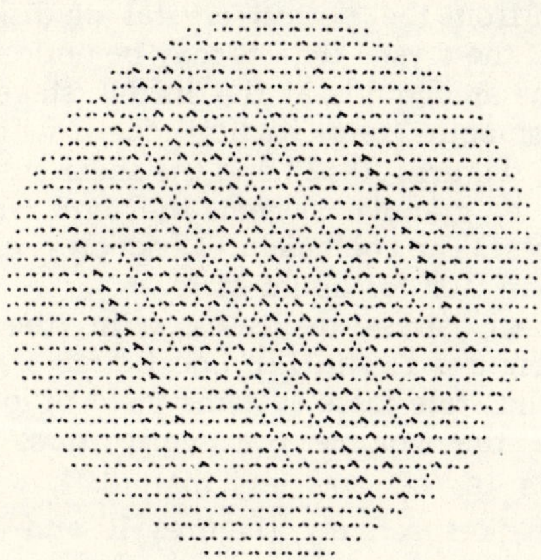

Figure 9. Spatial distributions of local instant strains at plastic deformation of perfect crystal of disks.

5. MECHANISM OF IRREVERSIBLE DEFORMATION (FLOW) IN LIQUID STATE.

Molecular theories of liquid flow suggest that the mechanism of flow is jump of a particle (or chain segments in polymers) into a hole [31,32,33]. However, molecular dynamics method [34] had revealed another flow mechanism. Its main feature is gradual displacement of neighboring particles positions.

As it was shown above irreversible deformation of a granular body in solid state is determined by disintegration of interparticle contacts under shear load. If granular system is in the liquid state number of contacts and density do not change during deformation [13-16]. The average number of contacts (per a particle) γ and the density ρ of two-dimensional (2D) system of smooth ellipses in liquid state are very high ($\gamma \approx 5.8$-6 and $\rho \approx 0.9$).

The goal of this chapter is a study of elliptic particles system flow mechanism.

Following 2D systems were studied [16]:

- Polydisperse system of randomly oriented and different sizes ellipses. A size distribution and an example of structure realization are shown in Figure 1.
- Perfect crystal of disks (Figure 2a).
- Monodisperse system of randomly oriented equal size ellipses (isotropic crystal) that were grown from the perfect crystal of disks (Figure 2b).

- Monodisperse system of unidirectionally oriented ellipses (anisotropic crystal that were grown from the perfect crystal of disks (Figure 2c). The spontaneous formation of the crystal defects may be noticed.

All these systems are liquids at the initial stage of deformation if elliptic particles are smooth (zero friction coefficient). The number of contacts does not change (Figures 10-12), and the shear stress is low.

In the literature on granular systems the term friction coefficient is used instead of shear stress. The coefficient of friction is equal to 0.5 - 0.6 for solid state and is less than 0.05 for liquid state.

In addition, shear stress decreases with the decrease of the calculation or test rate (Figures 11 and 12). Local strains are random (Figures 13-15). In contrast, in solid state these systems yield in local shear lines.

For liquid state the volume practically does not change during deformation. In all cases $|\varepsilon_V/\varepsilon_\tau|$ was less than 0.04, where ε_V and ε_τ are volume and shear strains respectively (Figures 10 and 16). In contrast, for solid state the volume of system under compression decreases. For Poisson's ratio of particles $\nu=1/3$ the ratio of volume to shear strains on the elastic part of stress- strain curve is estimated as $|\varepsilon_V/\varepsilon_\tau|\approx(1-\nu)/(1+\nu)\approx 0{,}5$. During yield in solid state the volume increases and $|\varepsilon_V/\varepsilon_\tau|$ is approximately 0.5 (Figure 16).

During deformation the particles turn [35,36] and the average orientation changes (Figure 17). As the orientation of the particles is initially isotropic the change of particle orientation results in change of the body shape, i.e. in shear deformation.

Let us name this mechanism of irreversible deformation of liquid a rotation flow. The following data allow to conclude that the rotation flow is the dominating mechanism of the system flow.

Figure 10 shows two different stress - strain curves for polydisperse system of ellipses. Curve 1 refers to the case when both translational and rotational motions of particles are allowed. Curve 2 is obtained for the frozen rotation of particles. Curves 1 and 2 reflect liquid and solid-like behavior respectively. The typical features of solid state are: comparatively high shear stress, disintegration of the interparticle contacts during shear deformation and change of the volume (on the initial elastic stage of deformation $|\varepsilon_V/\varepsilon_\tau|=0.5$; during yielding the system dilates and $|\varepsilon_V/\varepsilon_\tau|$ is approximately 0.5).

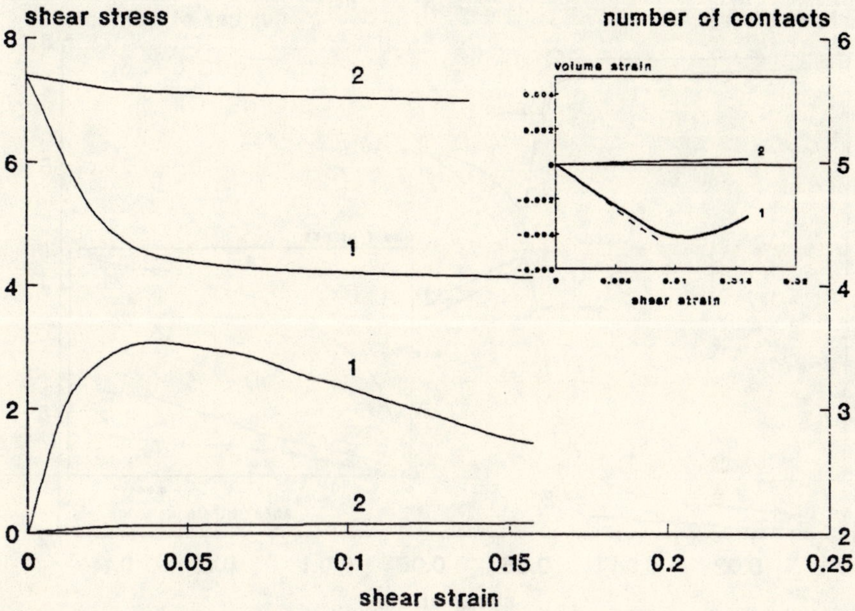

Figure 10. Biaxial compression ($\sigma_{11} > \sigma_{22}$ = const.) of polydisperse systems of smooth ellipses (e = 0,2) with frozen (1) and free (2) particle rotation; shear stress (a), number of contacts γ (b) and volume strain ε_V (c) versus shear strain ε_τ; σ_{22}≡2,0 (relative unit).

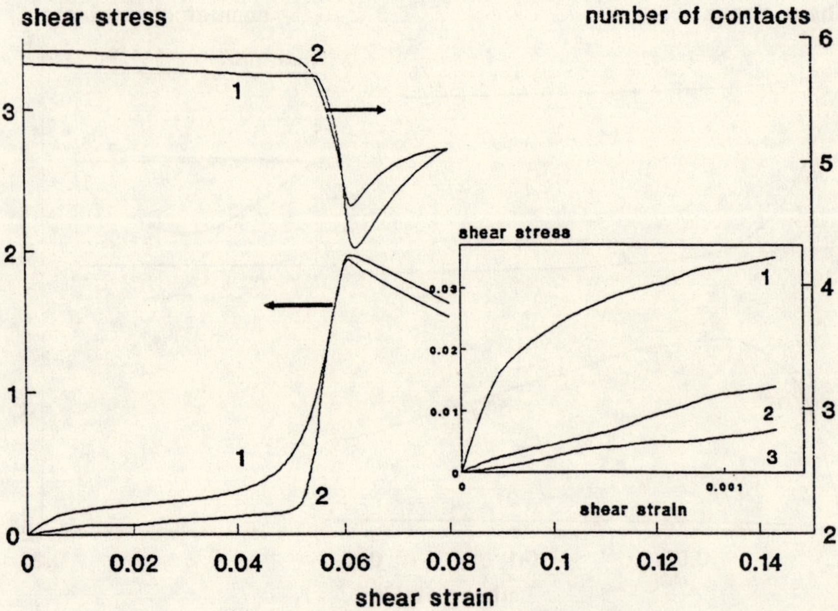

Figure 11. Biaxial compression ($\sigma_{11} > \sigma_{22}$ = const.) of monodisperse systems of smooth ellipses with e ≈ 0,027 (isotropic crystal); shear stress (a,b) and number of contacts γ (c) versus shear strain ε_τ; the ratio of strain rate is 1:0,1:0,01 for the curves 1,2 and 3 respectively.

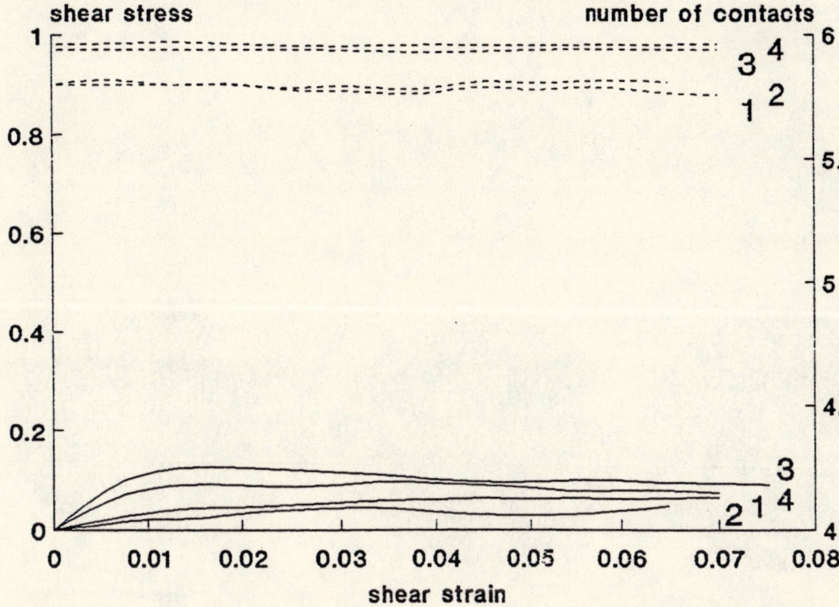

Figure 12. Biaxial compression ($\sigma_{11} > \sigma_{22}$ = const.) of monodisperse systems of smooth ellipses with e ≈ 0,213; isotropic crystal (1,2); anisotropic crystal (3,4); shear stress (a) and number of contacts γ (b) versus shear strain ε_τ; the ratio of strain rate is 1:0,1 for the curves 1,3 and 2,4 respectively.

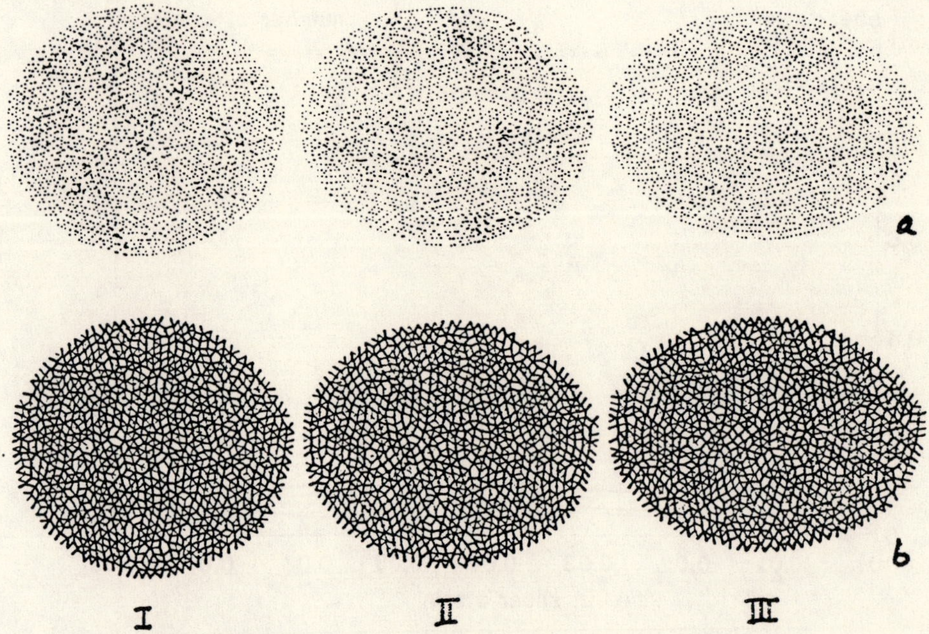

Figure 13. Distribution of local strains (a) and contact forces (b) under biaxial compression of system of smooth ellipses with different size (e = 0,2). The length of lines in Figure 13a is proportional to the displacement of a particle. The thickness of lines in Figure 13b is proportional to the contact force. Shear strain $\varepsilon_\tau$is equal to 0.05 (I), 0.1 (II) and 0.15 (III).

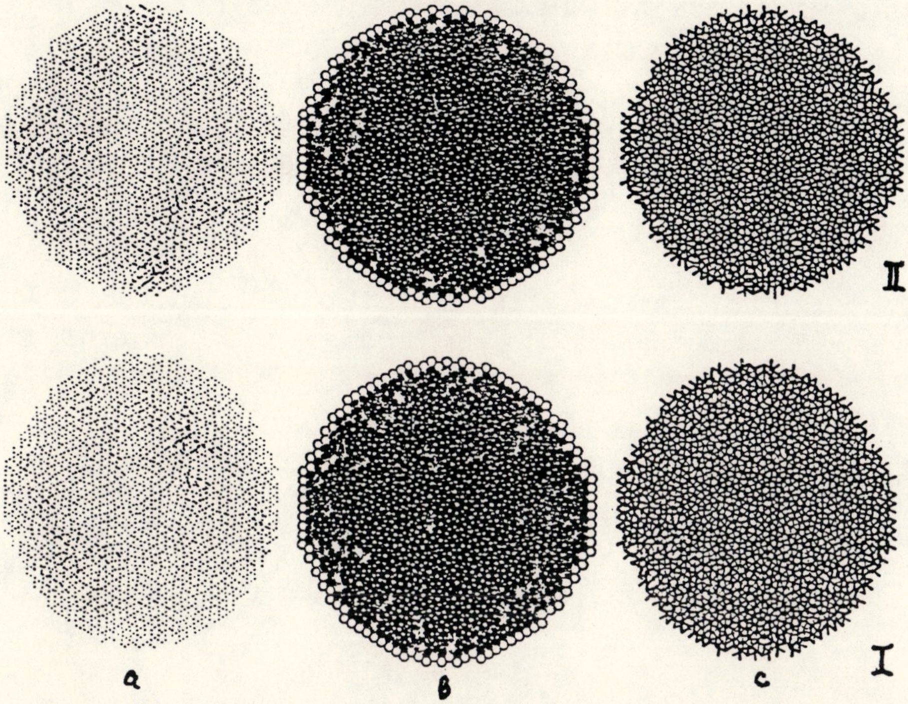

Figure 14. Biaxial compression of the isotropic system of smooth ellipses with e = 0.213. a- distribution of local shear strains. Number of contacts 6 (black) and <6 (white). b- distribution of local shear strains. The length of lines in Figure 14a is proportional to displacement of particles. The thickness of lines in Figure 14c is proportional to the contact force. Shear strain ε_τ=0.035 (I) and 0.06 (II).

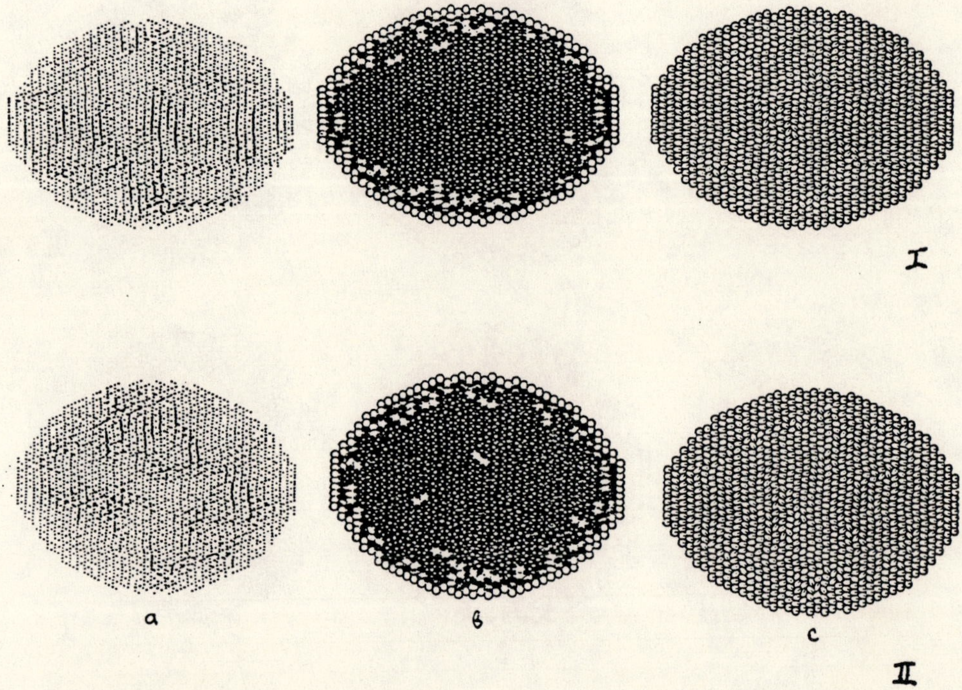

Figure 15. Biaxial compression of the anisotropic system of smooth ellipses with e = 0.213. a- distribution of local shear strains. b- distribution of particles with number of contacts 6 (black) and <6 (white). c- distribution of particle orientation. The length of lines in Figure 15a is proportional to displacement of particles. The thickness of lines in Figure 15c is proportional to the contact force. Shear strain ε_τ = 0,035 (I) and 0,07 (11).

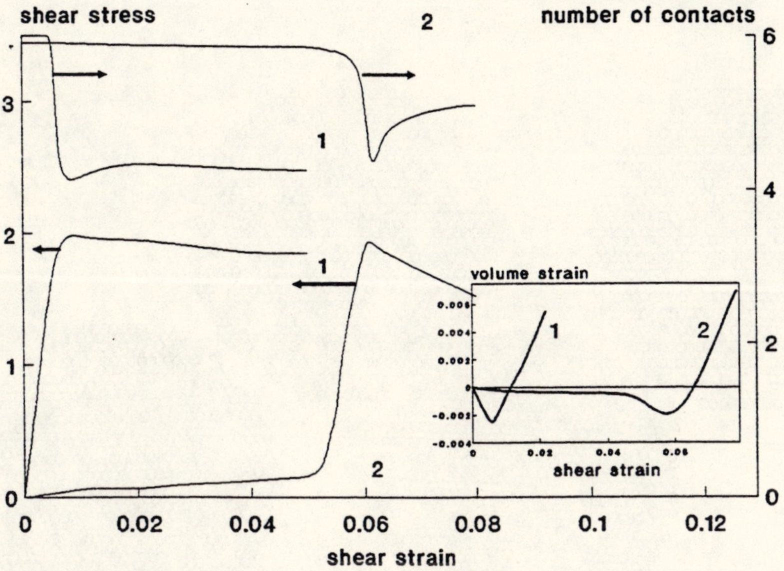

Figure 16. Biaxial compression ($\sigma_{11} > \sigma_{22}$ = const.) of perfect crystal of disks (1) and the isotropic system of smooth equal size ellipses with e ≅ 0,027 (isotropic crystal); shear stress (a); number of contacts γ (b); and volume strain ε_v (c) versus shear strain ε_τ.

This result contradicts the conclusion that granular materials are always liquid if the friction coefficient between particles is negligible [37].

Figures 17 and 18 show that the flow of initially isotropic crystal ellipses results in the orientation of the particles perpendicularly to the compression axis. At the initial part of deformation all features of the liquid behavior are observed: local strains are random (Figure 19), the shear stress is low, the number of contacts does not change, and the volume strain is very low (Figure 16).

Complete orientation of all particles causes the transition from liquid to solid state. The shear stress (friction coefficient) increases abruptly, the number of contacts γ decreases from 5.8-5.9 to ~4, and the shear deformation localizes in shear lines (Figure 9). The volume strain changes similarly to that in solid crystals or glasses (Figure 16c). Volume strain is initially negative (at compression) and later it is positive with high ratio of $|\varepsilon_v/\varepsilon_\tau|$=0.5.

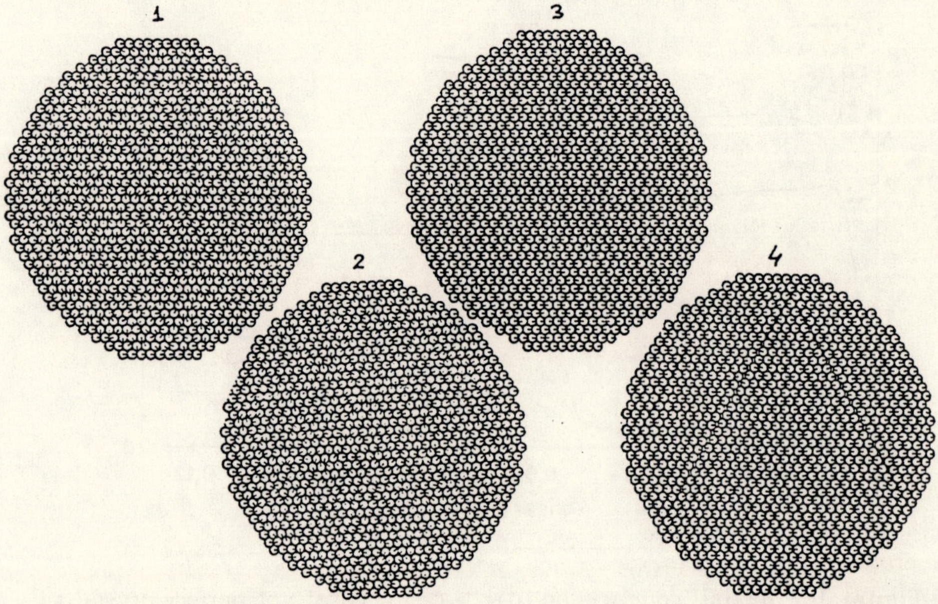

Figure 17. Change in the orientation of ellipses under compression in isotropic crystal of smooth ellipses with $e \approx 0.027$. Shear strain ε_τ is 0 (1), 0.02 (2), 0.065 (3) and 0.09 (4). Shear bands can be noticed in Figure 17-4.

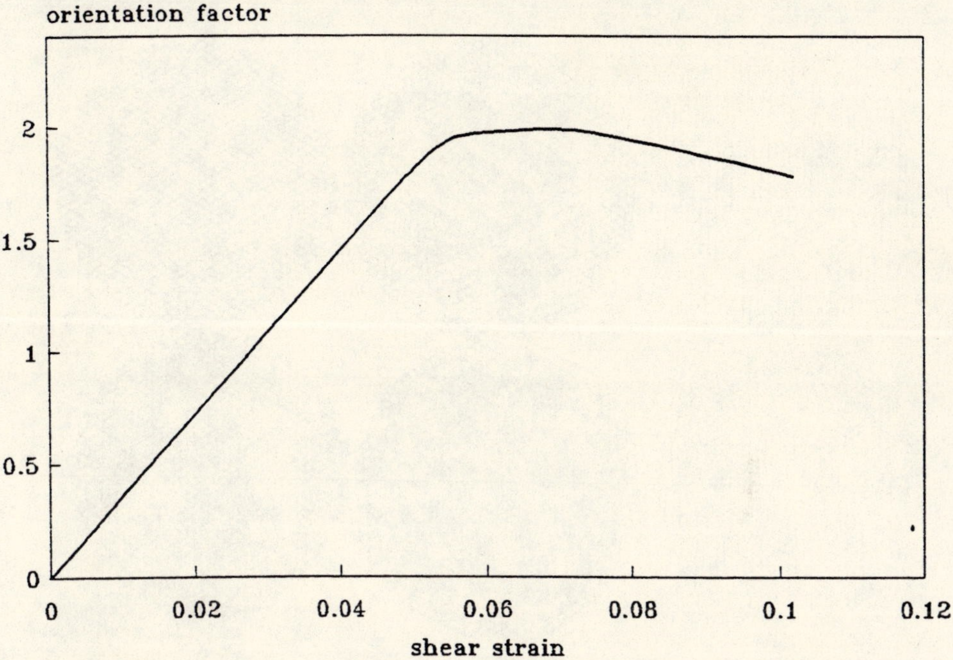

Figure 18. Ellipses orientation factor A_e versus shear strain ε_τ at compression of isotropic crystal of smooth ellipses with e≈0,027.

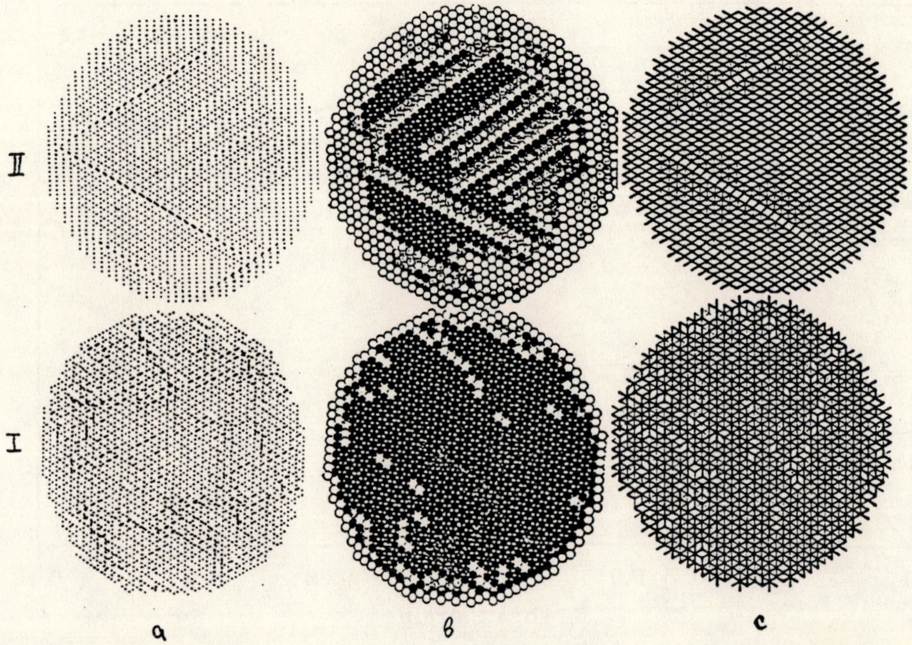

Figure 19. Biaxial compression ($\sigma_{11}>\sigma_{22}$ = const.) of isotropic crystal of smooth equal size ellipses with e ≈ 0.027 prior (I) and after complete orientation of ellipses. a-distributions of local strains. b-particles with number of contacts 6 (black) and <6 (white). c-distribution of contact forces. Average shear strain ε_τ is 0.02 (I) and 0.09 (II).

A simplified theory [16] describing the change in the irreversible shear strain and the average orientation of particles was developed. The theory is based on the assumptions: the volume does not change during deformation in liquid state. The length and the width are proportional to the sum of maximum sizes of particles in these directions. Entropy of particles is described by the equation: $S = {}_0\int^{\pi/2}\rho(\theta)\cdot\ln\{\rho(\theta)\}d\theta$, where $\rho(\theta)$ is the angle distribution of particles. Function $\rho(\theta)$ is calculated from the condition of entropy maximum at fixed body sizes. The particle rotation angle is related with shear deformation and the change in the body sizes (see Appendix).

The theoretical results qualitatively agree with the results of the computer calculations (Figure 20).

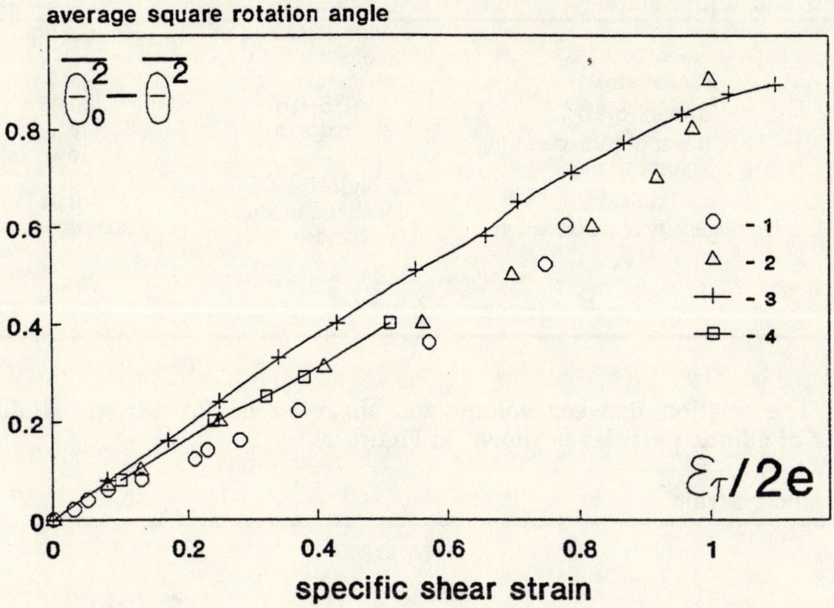

Figure 20. Average square angle of particles rotation, θ^2-θ_o^2, versus reduced shear strain, $\varepsilon_T/2e$; calculations by equation (4) (1 and 2) and computer experiments data (3 and 4) for isotropic systems of ellipses with e ≅ 0,027 (1 and 3) and e = 0,2 (2 and 4).

Thus the main mechanism of irreversible deformation of smooth ellipses in liquid state is the particles rotation. Rotation of ellipses leads to their orientation in the same direction and to the crystallization of the system. After this the system deforms as a solid body, and further irreversible deformation occurs only after disintegration of interparticle contacts.

The main features of irreversible deformation in solid and liquid state are summarized in Table 2.

Table 2. Comparative analysis of irreversible deformation parameters for solid and liquid state.

		Plastic deformation of solid	Flow of liquid
1	Shear stress $(\sigma_1-\sigma_2)/(\sigma_1+\sigma_2)$	high 0.5÷0.6	low <0.05
2	Number of contacts	reduces	constant
3	Volume change $\mid\varepsilon_v/\varepsilon_\tau\mid$	high ≅0.4÷0.5	low <0.04
4	Homogeneity of deformation	localized in shear bands	homogeneous

The relation between volume and shear strain for various studied systems of elliptic particles is shown in Figure 21.

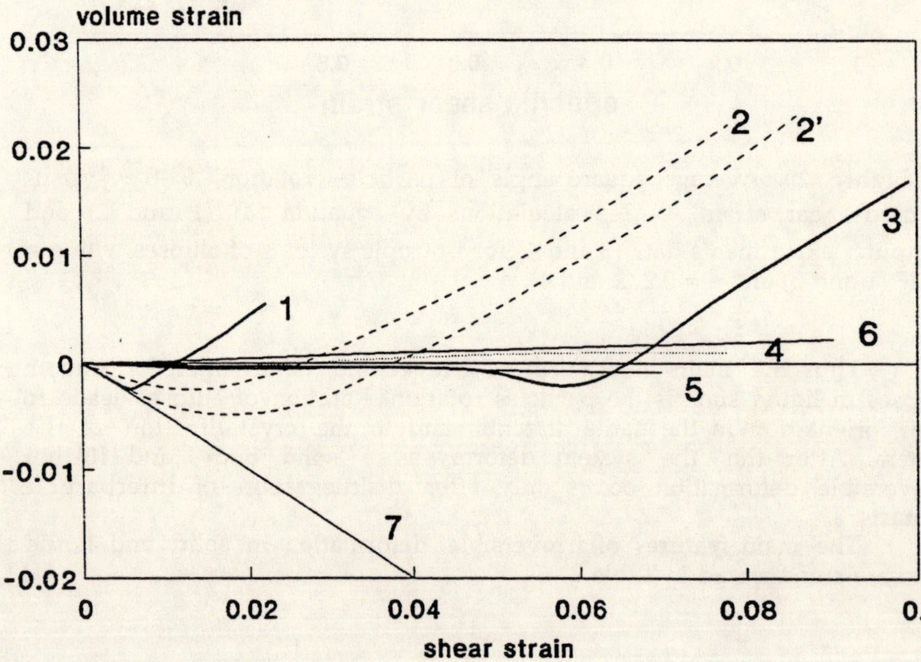

Figure 21. Volume strain versus shear strain at compression of various plane systems of hard elastic elliptic particles: 1 - perfect crystal of smooth disks; 2 - polydisperse disordered system of randomly oriented ellipses (e=0.2) with high interparticle friction; 2' - the same system of smooth ellipses with frozen particles rotation; 3 - system of randomly oriented equal

smooth ellipses with e=0.027; 4 - polydisperse disordered system of randomly oriented smooth ellipses with e=0.2; 5 - system of randomly oriented equal smooth ellipses with e=0.213; 6 - ordered system of uniformly oriented equal smooth ellipses with e=0.213 (crystal 1 with defects); 7 - elastic compression; the results of computer simulation.

6. MECHANICAL PROPERTIES OF PERFECT CRYSTALS OF RIGID SMOOTH EQUAL ELLIPSES, INTERMEDIATE ("LIQUID CRYSTAL") STATE.

In this chapter the detailed analysis of mechanical properties of various crystals of rigid smooth ellipses is carried out.

The elliptic particles are considered as completely rigid and smooth. In this case hydrostatic pressure was used for simulation of soft attractive branch of any interaction potential in a molecular system (for example, Leonard-Johns etc.). The density of perfect crystals of ellipses is numerically calculated by means of special program for a personal computer [38].

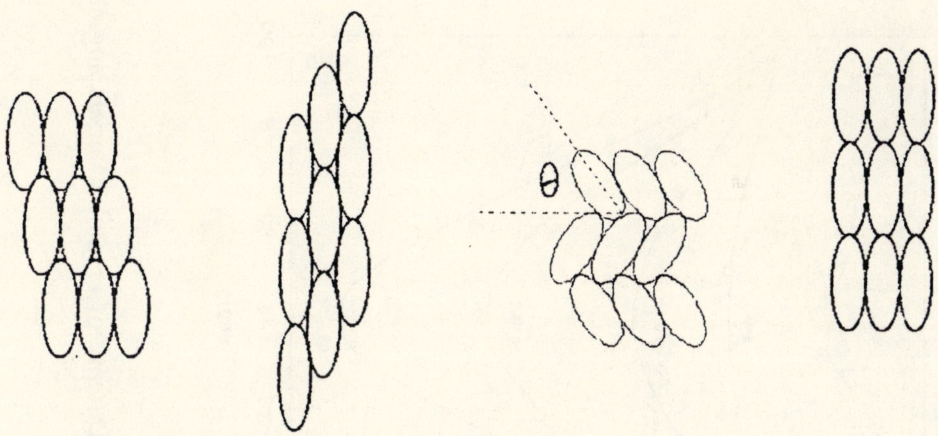

1 2 3 4

Figure 22. Structures of plane crystals of ellipses: 1, 2, 3 & 4.

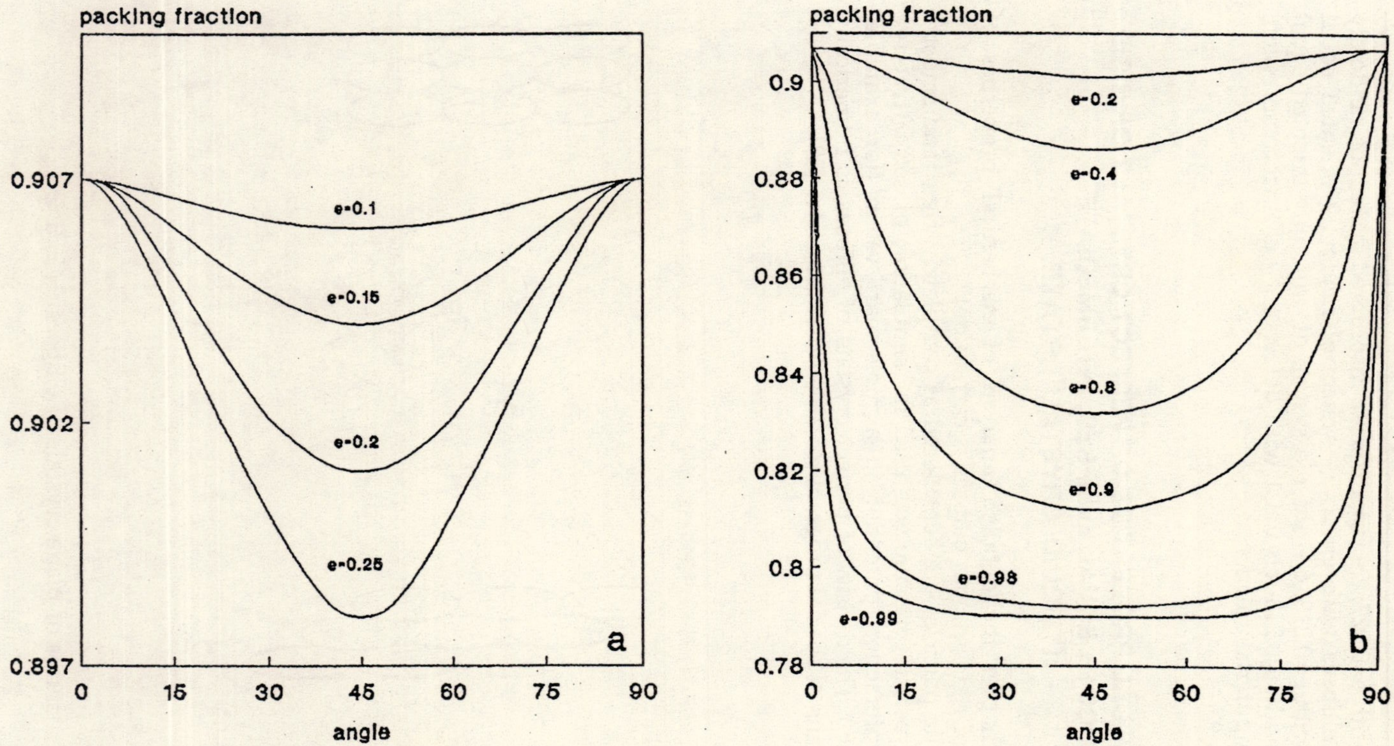

Figure 23. Density of crystal 3 versus angle between major axes of ellipses in neighbor layers of particles, θ.

In Figure 22 two types of the most dense crystals of ellipses, 1 and 2, and crystals with lower density, for example 3 and 4, the transformation of crystal 3 to 1 or 2 at θ - 0 or π/2 being observed . Density, ρ, of crystal 3 depends on θ (θ is an angle between major axes of ellipse particles in neighbor layers of crystal) and on ellipse eccentricity; densities of crystal 1 and 2 are the same and equal to density of the most dense packing of disks (hg) ($\rho = \pi/2\sqrt{3} \approx 0{,}9069$). Dependence ρ(θ) for crystal 3 is expressed by symmetrical curve which shows a minimum at θ = π/4 (Figure 23). Dependence of ρ(π/4) on ellipse eccentricity is shown in Figure 24. Crystals 1, 2 and 3 demonstrate maximum number of contacts, γ = 6.

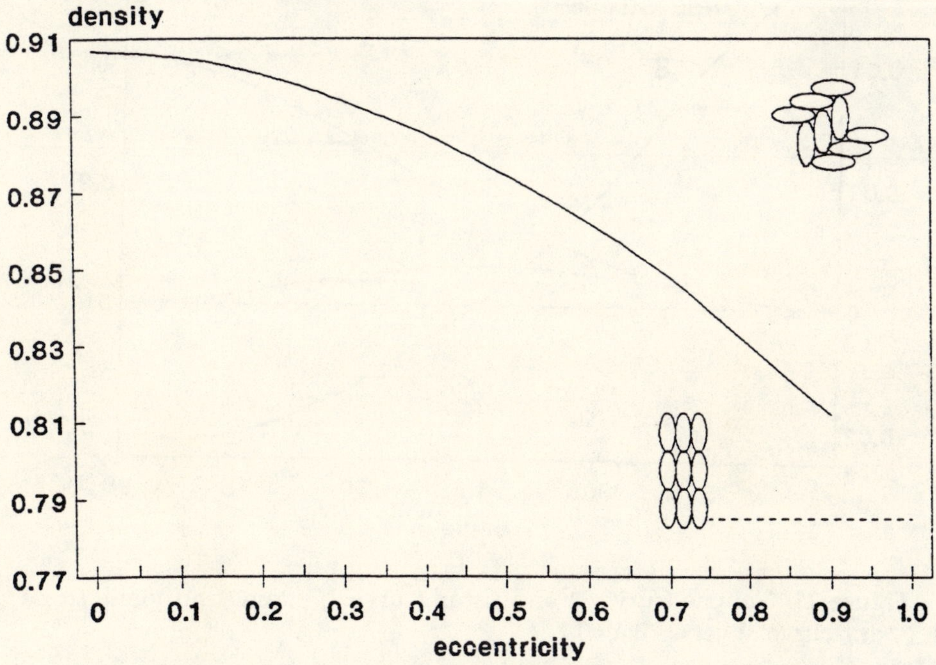

Figure 24. Density of crystal 3 with θ=π /4 versus ellipse eccentricity, e.

The change of volume having place in smooth rigid ellipses system under deformation is a result of changing: a) number of contacts, γ, and/or b) mutual particles orientation.

The role of the second factor may be elucidated by the study of transformation of crystals 1 and 2 to crystal 3 at biaxial compression ($\sigma_{11}>\sigma_{22}$) along major axes of ellipses. This is one of possible modes of homogeneous deformation at constant number of contacts γ ("liquid-like" flow). Just that very mechanism of deformation of ellipses crystal we observed in the computer experiment (see chapter 5), but the deformation in the sample was nonhomogeneous.

It should be noted that crystals 1 and 2 demonstrate very different responses to shear deformation regardless of the shape of the curve $\rho(\theta)$, whereas densities of the crystals are the same. But the curves of shear strain versus θ are not symmetrical and depend on ellipse eccentricity e: at $e \to 0$ the curve tends to symmetrical shape, and properties of crystals 1 and 2 become close (Figure 25).

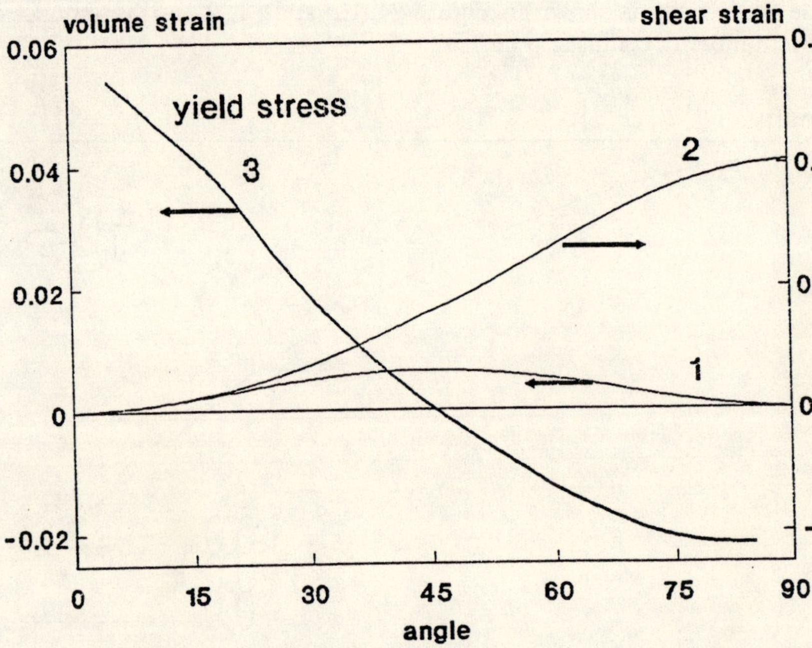

Figure 25. Volume (curve 1) and shear (curve 2) strains and yield stress ($d\varepsilon_V/d\varepsilon_\tau$, curve 3) versus θ; e=0.2.

The condition of body mechanical stability is a positive work of external forces, A ($A \geq 0$):

$$A=\int(\sigma_{11}\cdot d\varepsilon_{11}+\sigma_{22}\cdot d\varepsilon_{22})=\int P\cdot d\varepsilon_V+\int\tau\cdot d\varepsilon_\tau \text{ (at } \varepsilon_\tau<<1),$$
$$\text{where } P=(\sigma_{11}+\sigma_{22})/2 \text{ and } \tau=(\sigma_{11}-\sigma_{22})/2.$$

At shear deformation under biaxial compression ($\sigma_{11}<\sigma_{22}<0$) of expanding material ($d\varepsilon_V>0$, dilation) $\int P\cdot d\varepsilon_V<0$ and $\int\tau\cdot d\varepsilon_\tau>0$. Then the condition $A\geq 0$ is transformed to $\int\tau\cdot d\varepsilon_\tau \geq \int P\cdot d\varepsilon_V$.

In the case of body consisting of rigid smooth particles of arbitrary shape the elastic energy of particles remains constant (it is equal zero), and energy loss by friction is not observed, and hence the condition of equilibrium is dA=0. Then the equation for shear stress is deduced: $\tau = P \cdot d\varepsilon_V / d\varepsilon_\tau$, that corresponds to the shear stress - shear strain diagram (Figure 26). This shape of diagram is typical for buckling mechanism, for instance at compression of rigid rod. The maximum values of $|\tau|$ at $\theta=0$ and $\theta=\pi/2$ correspond to yield stresses for granular body (crystal 1 and 2 respectively).

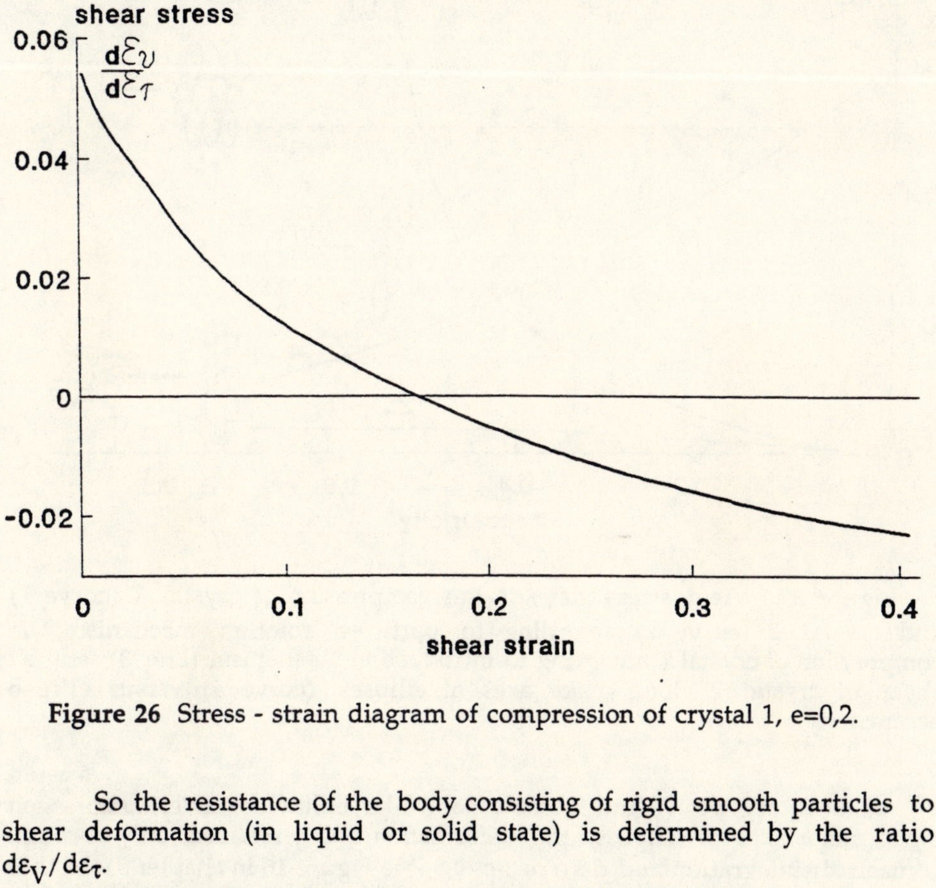

Figure 26 Stress - strain diagram of compression of crystal 1, e=0,2.

So the resistance of the body consisting of rigid smooth particles to shear deformation (in liquid or solid state) is determined by the ratio $d\varepsilon_V / d\varepsilon_\tau$.

The crystals 1 and 2 have maximum densities, and the volume is increased at shear deformation ($d\varepsilon_V > 0$, a dilatation). Dependence of $d\varepsilon_V / d\varepsilon_\tau$ on ellipse eccentricity e is shown in Figure 27 for two types of crystal, 1 and 2. The deformation of crystal 2 under shear stress is much higher than in the case of crystal 1 if ellipse eccentricity is sufficiently high and the compression is directed along major axes of ellipses. Both crystals show liquid like behavior at e→O ($d\varepsilon_V/d\varepsilon_\tau \to 0$). Crystal 2 retains liquid (or quasi-liquid) character at any eccentricity ($d\varepsilon_V/d\varepsilon_\tau < 0,04$), while crystal 1 becomes solid ($d\varepsilon_V/d\varepsilon_\tau > 0,5$) at high eccentricity.

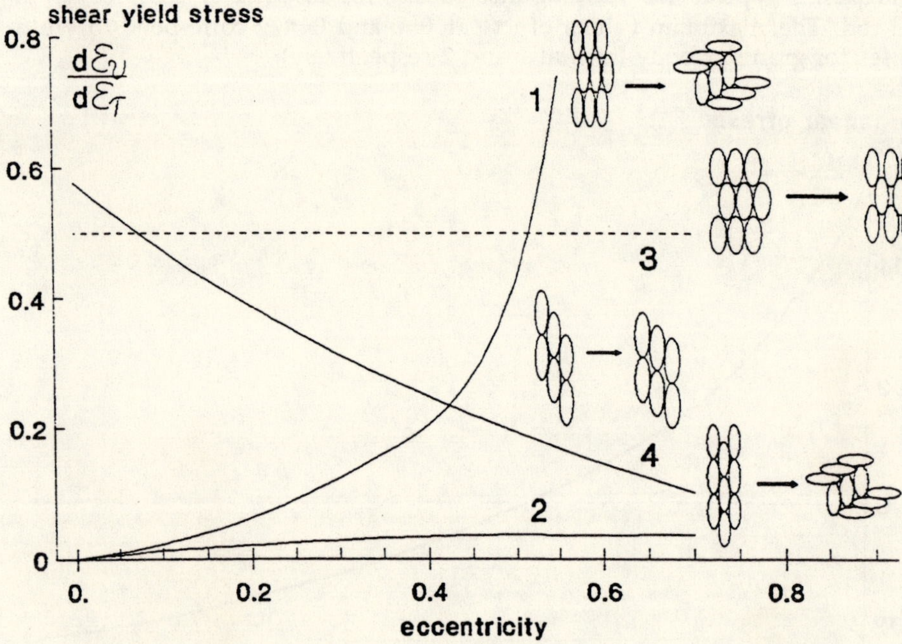

Figure 27. Yield stress ($d\varepsilon_V/d\varepsilon_\tau$) at compression of crystal 1 (curve 1) and crystal 2 (curve 2) according to particles rotation mechanism; at compression of crystal 1 according to introduction mechanism (line 3); and at shear of crystal 2 along major axes of ellipses (curve 4) versus ellipse eccentricity.

Both crystals (1 and 2) show solid-like behavior under compression along minor axes of ellipses, the deformation being realized only through contacts disintegration and $d\varepsilon_V/d\varepsilon_\tau$ -> 0,5 (see Figure 16 in chapter 5). In this case the strain localization in shear bands is observed (Figure 19 in chapter 5).

In the case of introduction of one layer of crystal 1 in neighbor layer and transformation of crystal 1 to crystal 4, the number of contacts being reduced from 6 to 4 (γ→4), the calculation shows that $d\varepsilon_V/d\varepsilon_\tau = 0,5$ (Figure 27). The value of $d\varepsilon_V/d\varepsilon_\tau$ 0,5 was observed in various cases of plastic deformation of solids such as perfect crystal of disks, anisotropic crystal of ellipses under compression along minor axes of ellipses, glass of polydisperse ellipses at high friction or frozen particles rotation (Figure 21).

Thus, the change in deformation mechanism in the case of compression of crystal 1 along major axes of ellipses is observed at $e \approx 0{,}53$, i.e. the transition from mechanism a (the transformation of crystal 1 to 3 by particles rotation with constant number of contacts at $e<0{,}53$) to mechanism b (the introduction of one layer of crystal 1 in other layer with disintegration of interparticle contacts (the and the transformation of crystal 1 to 4 at $e > 0{,}53$) (Figure 27).

Breaking of interparticle contacts at the shear of crystal 2 along major axes of ellipses ($\gamma \to 4$) also takes place, and $d\varepsilon_V/d\varepsilon_\tau$ $(1-e)/\sqrt{3}\cdot(1+e)$. In this case the crystal 2 becomes "liquid-like" only at high eccentricity of ellipse ($e \to 1$), but shear resistance of crystal 2 remains sufficiently high for ellipses with $e<0.7$ since $d\varepsilon_V/d\varepsilon_\tau > 0{,}1$ (Figure 27).

The analysis of deformation process for perfect crystals is described above. Indeed, various crystal defects are of importance in the process of deformation of real crystal bodies, especially solids. In particular, we observed crystal defects (inclusions of structure 3 in crystal 1) in computer experiment on the growth of ellipses crystal from perfect crystal of disks at gradual increase of eccentricity (Figure 2c). In fact such defects reduce the yield stress of material.

Thus, in systems of equal ellipses two the most comfortable (the most dense) states exist (crystals 1 and 2). These structures are anisotropic and have different mechanical properties. Polycrystal of equal ellipses may be a mixture of small crystals and 2. In the process of polycrystal deformation a jump or gradual transition from one structure to another may be observed. In this case the final deformation takes place at practically constant volume because the densities of structures 1 and 2 are the same. Then the state 3 is a defect, and total alteration of body volume depends on the number and density of defects.

The above considerations demonstrate that the resistance of crystals of smooth rigid elliptic particles to shape change (yield stress, τ_y) in conditions of static equilibrium (that corresponds to zero temperature, O K, for molecular substances) is varied from zero ($\tau_y=0$, liquid state) to the value that is typical for the solid of ideal structure ($\tau_y \approx 0{,}5\cdot P$), intermediate values being observed depending on the shape (eccentricity) and mutual disposition (angle θ) of particles in crystal. Hence, the existence of crystal structures with very low yield stress ($\tau_y \to 0$) - liquid (or quasi-liquid) crystal is possible in principle even at $T \to O$. Here the definition of crystal is related to the structure of material - a presence of long-range order of particles location and orientation, and the definition of liquid is related to the mechanics of material - an ability to the shape change under negligibly low (or extremely low) shear stress.

7. CONCLUSIONS

In this paper we have tried to reveal the basic behavior characteristics of plane (2D) bodies consisting of hard particles of various shape under the action of external mechanical forces. The results obtained are strictly valid for granulated materials due to existence of high repulsive potential (up to "hard wall").

The potential of any interatomic or intermolecular interaction includes sufficiently hard repulsive branch that permits to consider the possibility of basic qualitative conclusions described in this paper being valid for molecular systems.

The features of computer experiment in this case make it possible to relate characteristics studied to the conditions of quasi-static equilibrium, i.e. to extremely slow loading ($\varepsilon \rightarrow 0$, where ε is deformation rate) in the absence of heat motion of particles ($T \rightarrow O$ K).

It should be also emphasized that studied systems being under conditions of quasi-static mechanical equilibrium (local energy minimum) are not always at thermodynamic equilibrium that is characterized by global energy minimum. Thus, the mixture of crystals each consisting of equal particles corresponds to global minimum energy for infinite system of rigid particles of various sizes being under overall pressure. In the case of rigid particles system under pressure energy minimum corresponds to the maximum packing fraction (density). The crystal of equal rigid ellipses with an arbitrary angle θ between neighbor layers of particles (crystal 3) and isotropic system of equal ellipses of random orientation (Figure 2b) do not correspond to global minimum, i.e. total thermodynamic equilibrium as well.

This study does not concern the causes of systems formation and stabilization of some initial states (local energy minimum) but about these states transformation under external load.

The following results seem to be of most great interest:

1. The disordered (amorphous) bodies consisting of various size particles may be (under above mentioned conditions) in solid (glass) or liquid state in dependence on the number of interparticle contacts, γ: the condition of $\gamma > \gamma^*$ corresponds to glass, that of $\gamma < \gamma^*$ corresponds to liquid, where γ^* is the critical number of contacts.

2. The critical number of contacts, γ^*, as well as the density (packing fraction) for a system being at the point of liquid - glass transition depends on symmetry of particle shape and con tact force. It changes within rather wide range that, firstly, is contrary to the free volume models and, secondly, may describe glass transition as a second-order phase transition.

3. The development of irreversible deformation in solid state (plasticity) in disordered (glass) as well as ordered (perfect crystal) systems requires the exceeding of some level of shear stress (yield stress). This deformation is determined by disintegration of interparticle contacts and is accompanied by the reduce of packing fraction (dilatation) and localization of deformation in the form of shear bands.

In the presence of defects and non-uniformity of structure, e.g., dislocation in crystals, essential change of mechanism and parameters of plastic deformation may be observed.

4. The irreversible deformation in disordered systems of smooth ellipse in liquid state (flow process) is determined by particles rotation and changes in the orientation, the number of contacts and the packing fraction being preserved, and local strain distribution being random.

This flow mechanism is typical for dense granular systems ($\gamma \approx \gamma^*$) whereas other mechanisms are also possible for loose systems ($\gamma < \gamma^*$).

5. The resistance of crystal structures of equal smooth rigid ellipses to shape change (yield stress, τ_y) under conditions of static equilibrium may change from zero ($\tau_y = 0$, liquid state) to the typical value for solid bodies of ideal structure ($\tau_y \approx 0{,}5$-P). Intermediate values are observed depending on particle shape (the eccentricity), their mutual arrangement in crystal (θ angle) and loading direction.

The deformation of anisotropic crystal of unidirectional ellipses may correspond to the case of liquid at compression along major axis of ellipses and to the case of solid body at compression along minor axis.

The deformation of isotropic crystal is typical for a liquid until complete ellipses orientation with subsequent system crystallization. Then the body becomes solid in the direction of deformation, and subsequent irreversible shear deformation in the same direction can be realized by contact disintegration only.

It means that crystal structures with extremely low yield stress ($\tau_y \to 0$), i.e. liquid (or-quasi-liquid) crystals may exist in principle at $T \to 0$. In this case the notion of crystal is of structural character (long-range order in particles arrangement and orientation), and the notion of liquid is a mechanical one (body ability to shape change under negligibly low (or extremely low) shear stress.

6. In all case of irreversible deformation in solid as well as liquid state no particle jumps to neighbor void (vacancy mechanism) are, but only gradual relative displacement of the particles has place. It makes necessary the use of critical approach to the Eyring - Frenkel and free volume models, etc.

The subsequent development of the approach described in this work will be aimed to the elucidation of the influence of inter action potential shape (particularly at the transition from hard to more soft potential) and kinetic energy (temperature) on the conditions of liquid - glass transition and mechanical characteristics of examined models.

8. ACKNOWLEDGMENTS

We are very grateful to Professors L.I. Manevitch and V.G. Oshmyan for useful discussions.

9. APPENDIX

Let us consider the case of uniaxial tension - compression of elliptical particles of the same size supposing that the body size in the direction of tension - compression is proportional to a sum of ellipses maximum sizes in this direction:

$$L \sim \int_0^{\pi/2} l(\theta) \cdot \rho(\theta) d\theta$$

where $l(\theta)$ the height of ellipse with inclination angle θ to the tension axis, and $\rho(\theta)$ - the distribution function of particles orientation. If e - the eccentricity and d - the average size of ellipse, $l(\theta)$ may be expressed as follows:

$$l(\theta)/d = \sqrt{1+e^2+2e \cdot \cos 2\theta}$$

Taking into account practically constant volume of system at deformation in liquid state, we may write (at $\varepsilon_\tau << 1$):

$$\varepsilon_\tau/2 = \varepsilon_{11} = -\varepsilon_{22} = (L-L_0)/L_0. \quad (1)$$

Let us introduce the orientation entropy to determinate the function $\rho(\theta)$:

$$S = \int_0^{\pi/2} \rho(\theta) \cdot \ln\{\rho(\theta)\} d\theta$$

and let us determinate entropy maximum at constant body shape:

$$L = \text{const.}$$

Then we obtain:

$$\rho(\theta) = C_1 \cdot exp[-\mu \cdot 1(\theta)]$$

where $C_1 = \{\int_0^{\pi/2} exp[-\mu \cdot 1(\theta)] \cdot d\theta\}^{-1}$ - normalization factor, μ - a parameter of distribution of ellipses orientation.

$$L \sim C_1 \cdot {}_o\int^{\pi/2} l(\theta) \cdot exp[-\mu \cdot l(\theta)] \cdot d\theta$$

or:

$$L \sim C_1 \cdot {}_o\int^{\pi/2} \sqrt{1+e^2+2e \cdot \cos 2\theta} \cdot exp(-\mu \cdot \int 1+e^2+2e \cdot \cos 2\theta) \cdot d\theta \qquad (2)$$

A parameter of ellipses orientation, A_e, included in interpolation formula for a distribution function of particles orientation, $\rho(\theta) = 2/\pi \cdot (1+A_e \cos 2\theta)$, was calculated using method of least squares.

Average angles of ellipses orientation in relation to the loading direction may also be calculated:

$$\theta = C_1 \cdot {}_o\int^{\pi/2} \theta \cdot \exp[-\mu \cdot l(\theta)] \cdot d\theta \qquad (3)$$

$$\theta^2 = \{C_1 \cdot {}_o\int^{\rho/2} \theta^2 \cdot \exp[-\mu \cdot l(\theta)] \cdot d\theta\}^{1/2} \qquad (4)$$

The analysis of initial stages of deformation shows that following relationships are valid:

$$(dA_e/d\varepsilon_\tau) \cdot e = (d\theta/d\varepsilon_\tau) \cdot e = (d\theta^2/d\varepsilon_\tau) \cdot e = \text{const.}$$

Numerical calculations result to following values:

$$(\frac{\varepsilon_\tau}{\mu e^2})_{\mu \to 0} = 1, \ (\frac{dA_e}{d\varepsilon_\tau} \bullet e)_{\mu \to 0} \cong 1{,}0;$$

$$(\frac{d\bar{\theta}}{d\varepsilon_\tau} \bullet e)_{\mu \to 0} \cong 0{,}32; \ (\frac{d\bar{\theta}^2}{d\varepsilon_\tau} \bullet e)_{\mu \to 0} \cong 0{,}28;$$

Thus, in equations (1), (2), (3) and (4) the relationship between strain and degree of ellipses orientation is established. The numerical solution of these equations are shown in Figure 20.

10. REFERENCES

1. Cundall P.A. and Strack O.D.L. *1979 Geotechnique* 29 47.

2. Rothenburg L *1980 Micromechanics of Idealized Granular Systems*, Ph.D. Dissertation, Carlton University, Ottawa, Ontario, Canada.

3. Bathurst R J 1985 *A Study of Stress and Anisotropy in Idealized Granular Assemblies*, Ph.D. Dissertation, Queens University at Kingston, Ontario, Canada.

4. Thornton C and Barnes D J *1986 Acta Mechanica* 64 45.

5. Rothenburg RJ and Bathurst R J *1989 Geotechnique* 39 601.

6. Bathurst R J and Rothenburg L 1988 *J. Appl. Mech.*, ASME 55 17.

7. Bathurst R J and Rothenburg L 1988 *Int. J. Eng. Sci.* 26 373.

8. Bathurst R J and Rothenburg L 1990 *Mechanics of Materials* 9 65.

9. Rothenburg L, Bathurst R J and Dusseault M 1989 *Powder and Grains*, ISBN 355.

10. Rothenburg L and Bathurst R J *1991 Computers and Geotechnics* 11 315

11. Rothenburg L and Bathurst R J *1992 Geotechnique* 42 79.

12. Edwards S F *1990 Rheologica Acta* 29 493.

13. Berlin Al Al, Rothenburg L and Bathurst R J *1991 Khim. Phyz.* 10 1284 (in Russian).

14. Berlin Al Al, Rothenburg L and Bathurst R J *1992 Vysokomol. soed. A* 34 6, 1992 Polymer Science USSR 34 559.

15. Berlin Al Al, Rothenburg L and Bathurst R J *1993 Vysokomol. soed. A* 35 857, 1993 Polymer Science USSR 35 957.

16. Berlin Al Al, Rothenburg L and Bathurst R J *1993 Vysokomol. soed. A* 35 864, 1993 Polymer Science USSR 35 957.

17. Ottavi H, Clere J, Girand G, Roussenq J, Guyon E and Mitescu C D 1978 J. *Phys. C: Solid State Phys.* 11 1311.

18. Phillips J C February 1982 *Physics Today* 27.

19. Phillips J C *1980 Comments on Solid State Phys.* 9 191.

20. Phillips J C *1981 Comments on Solid State Phys.* 10 85.

21. Berlin A A, Korolev G V, Kefeli T Ya and Sivergin Yu M *1983 Acrylic Oligomers and Related Materials* (Moscow: Khimiya) (in Russian).

22. Yu C C and Leggett A J *1972 Comments Cond. Mat. Phys.* 14 231.

23. Anderson P W, Halperin B I and Varma C M *1972 Philos. Mag.* 25 1.

24. Phillips W A 1972 *J. Low Temp. Phys. Rev. B* 7 351.

25. Gibbs J H and DiMarzio E A 1958 *J.Chem. Phys.* 28 373.

26. Gilman J J 1975 *J.Appl.Ph.* 46 1625.

27. Li J M 1978 In *Metallic Glasses*, ed J J Gilman (Ohio: Metals Park) p.173.

28. Spaepen F 1977 *Acta Metall.* 25 407.

29. Argon A S 1979 *Acta Metall.* 27 47.

30. Grineva N S, Karpenko Yu P, Berlin Al Al and Manevich L I 1982 *Dokl. Acad. Nauk. SSSR* 266 384 (in Russian).

31. Frenkel Ya I 1953 Proceedings vol. 3 *Kineticheskaya teoriya zidkostey* (Moscow-Leningrad: USSR Academy of Sciences).

32. John M C, Klotz W L and Eyring H 1969 *J. Chem. Phys.* 51 3692

33. Vinogradov G V and Malkin A Ya 1980 *Rheology of Polymers* (Moscow: Mir) (in Russian).

34. Alder B J and Weinwright T E 1970 *J. Chem. Phys.* 53 3813

35. Konishi J, Oda M and Nemat-Nasser S -1983 *Mechanics of Granular Materials: New Models and Constitutive Relations*, ed J T Jenkins and M Satake (Amsterdam: Elsevier) p.31.

36. Oda M 1972 *Soil and Foundations* 12 1.

37. Oda M, Konishi J and Nemat-Nasser S 1983 *Mechanics of Granular Materials: New Models and Constitutive Relations*, ed J T Jenkins and M Satake (Amsterdam: Elsevier) p.21.

38. Berlin Al Al, Grigoryan G A and Rothenburg L 1993 *Vysokomol. soed. B* 35 1553, 1553 *Polymer Science USSR* 35 1297.

Subject Index

S

T

V